RENEWALS 458-4574

DATE DUE

GAYLORD			PRINTED IN U.S.A.

Technology Transfer and Industrial Change in Europe

Also by Helen Lawton Smith

ENERGY AND ENVIRONMENT REGULATION (*editor with Nick Woodward*)

ENVIRONMENTAL REGULATION IN TRANSFORMING ECONOMIES: The Case of Poland (*editor with Piotr Jasinski*)

PRIVATIZATION AND DEREGULATION OF TRANSPORT (*editor with Bill Bradshaw*)

Technology Transfer and Industrial Change in Europe

Helen Lawton Smith
Reader in Local Economic Development
Centre for Local Economic Development
Coventry Business School
Coventry University
and
Senior Research Associate
School of Geography
University of Oxford

First published in Great Britain 2000 by
MACMILLAN PRESS LTD
Houndmills, Basingstoke, Hampshire RG21 6XS and London
Companies and representatives throughout the world

A catalogue record for this book is available from the British Library.

ISBN 0–333–60458–X

First published in the United States of America 2000 by
ST. MARTIN'S PRESS, INC.,
Scholarly and Reference Division,
175 Fifth Avenue, New York, N.Y. 10010

ISBN 0–312–23110–5

Library of Congress Cataloging-in-Publication Data
Lawton Smith, Helen
Technology transfer and industrial change in Europe / Helen Lawton Smith
p. cm.
Includes bibliographical references and index.
ISBN 0–312–23110–5 (cloth)
1. Technology transfer—Europe. 2. Research, Industrial—Europe.
I. Title

T174.3 L38 2000
338.9406—dc21
                                                             99–053642

This book is printed on paper suitable for recycling and made from fully managed and sustained forest sources.

10   9   8   7   6   5   4   3   2   1
09   08   07   06   05   04   03   02   01   00

Printed and bound in Great Britain by
Antony Rowe Ltd, Chippenham, Wiltshire

Contents

v

List of Tables

List of Figures

Preface and Acknowledgements

This book is based on a two-year study (October 1990–September 1992) funded by the Joint Committee of the ESRC and SERC entitled 'Technology Transfer in the UK and European Context: the Flow Measuring Industry and the Electronic Component Industry' (award number GR-9198282). The principal investigator was Erik Swyngedouw. The study focused on the extent and nature of links which firms in the two sectors have with universities and national laboratories. The experiences of industry and researchers in the public sector, of government departments which provide funding for collaborative projects and of industry organizations are recorded in this book. The project also sought, particularly through Erik Swyngedouw's contribution, to develop theoretical insights into what is an increasing but by no means universal trend – that of externalization of innovation.

Special thanks are due to the project steering group, listed in Appendix 1, drawn from both industry and academia, who provided an invaluable contribution to the project, acting as a sounding board and giving practical advice on many issues. I am also grateful to Erik Swyngedouw and Gordon Clark for providing many useful suggestions on this work, and to Rupert Waters of the Centre for Local Economic Development, Coventry Business School, Coventry University, for proof reading and commenting on the manuscript. I would also like to thank Sunder Katwala, then at the Macmillan Press, for his helpful advice and Alison Howson, his successor, for seeing the project through. Janey Fisher, the freelance editorial services consultant on this book, is thanked for her efficiency and kindness in editing and preparing the manuscript for production. Finally, I would like to acknowledge the permission of UCL Press (now Taylor, Francis Routledge) to reprint Erik Swyngedouw, Martine Lemattre and Peter Wells' map of regional distribution of employment in electronic components, 1981.

List of Abbreviations

ABRC	Advisory Board for the Research Councils (UK)
ACARD	Advisory Council
ACOST	Advisory Council on Science and Technology (UK)
ANVAR	Agency Nationale de Valorisation de la Recherche (France)
ASICs	Application Specific Integrated Circuits
BCR	Community Bureau Reference (EU)
BRITE	Basic Research in Industrial technologies for Europe (EU)
CASE	Cooperative Awards in Science and Engineering (UK)
CBI	Confederation of British Industry (UK)
CEA	Commissariat à l'énergie atomique (France)
CERN	European Organization for Nuclear Research (Geneva)
CERT	Centre d'Etudes et de Recherches de Toulouse (France)
CIFRE	Conventions Industrielles de Formation par la Recherche (France)
CMOS	Complementary Metal Oxide Semiconductor
CNES	Centre national d'études spatiales (France)
CNRS	Centre National de la Recherché Scientifique (France)
CNET	Centre national d'études des Télécommunications (France)
CRITT	Centres régionaux d'innovation et de transfert de technologies
DATAR	Délégation à l'Aménagment du Territoire et à l'action régionale (France)
DERA	Defence Evaluation and Research Agency (UK)
DRA	Defence Research Agency (UK)
DRAMs	Dynamic Random Access Memories
DRRT	Délégations Régionales de la Recherche et de la Technologie (France)
DTI	Department of Trade and Industry (UK)
EC	European Commission
ECU	European Currency Unit
EM	Electro-magnetic
EPROMS	Electrically Erasable Programmable Read Only Memories
EPSRC	Engineering and Physical Sciences Research Council (UK)
ESA	European Space Agency

ESPRIT	European Strategic Programme of Research in Information Technology
ESRC	Economic and Social Research Council (UK)
EXERA	Association des Exploitants d'Equipements de Mesure, de Regulation et d'Automatisme (France)
EU	European Union
EURAM	European Research in Advanced Materials (EU)
FLOMIC	Flow Measurement Instrumentation Consortium (UK)
GDP	Gross Domestic Product
GTT	Groupe De Tranferet De Technologie (Belgium)
HEFC	Higher Education Funding Council (UK)
HEIs	Higher Education Institutes
ICSE	Integrated Control Systems Engineering Committee (UK)
IMEC	Inter-university Microelectronics Centre (Belgium)
INRIA	National Institute for Research in Information Technology and Automation (France)
IPR	Intellectual Property Rights
ISHM	International Society for Hybrid Micro-electronics
IWT	Flemish Research and Technological Development Fund (Belgium)
JESSI	Joint European Submicron Silicon Initiative (EU)
JFIT	Joint Framework on Information Technology (UK)
JIMS	Joint Industrial Measurements Systems Programme (UK)
JOERS	Joint Opto Electronics Research Scheme (UK)
KUL	Katholieke Universiteit te Leuven (Belgium)
LETI	Laboratoire D'electronique, de Technologie et d'Instrumentation (France)
MDSPSRE	Multi Departmental Scrutiny of Public Sector Research Establishments (UK)
MECU	Million European Currency Units
MEN	Ministry of National Education (France)
MIT	Massachusetts Institute of Technology (USA)
MNCs	Multi-national Corporations
MoD	Ministry of Defence (UK)
MPC	Multi-Project Chip Service (Belgium)
MRT	Ministry of Research and Technology (France)
NAMAS	National Measurement Accreditation Service (UK)
NAP	Nomenclature des Activites et des Produits (France)
NEDC	National Economic Development Council (UK)
NEL	National Engineering Laboratory (UK)
NSF	National Science Foundation (Belgium)

OECD	Organization of Economic Cooperation and Development
ONERA	Office National d'Etudes et de Recherches Aerospatiales (France)
OSCAR	Optical Sensors Club (UK)
OST	Office of Science and Technology (UK)
PCBs	Printed Circuit Boards
PCFC	Polytechnic and Colleges Funding Council (UK)
PCIF	Printed Circuit Board Federation (UK)
POST	Parliamentary Office of Science and Technology (UK)
PSSB	Public Sector Science Base
RACE	Research and Development in Advanced communications Technologies in Europe (EU)
R&D	Research and Development
RTD	Research and Technological Development
RUG	Rijksuniversiteit te Gent (Belgium)
SBS	Save British Science (UK)
SCK	Studiecentrum Voor Kernenergie (Belgium)
SERC	Science and Engineering Research Council (UK)
SERV	Sociaal-Economische Raad Van Vlaanderen (Belgium)
SIC	Standard Industrial Classification (UK)
SIREP	SIRA Instrumentation Panel (UK)
SMEs	Small and Medium Sized Enterprises
SPRU	Science Policy Research Unit (Sussex University) (UK)
THES	Times Higher Education Supplement
TNCs	Transnational Corporations
UFC	Universities Funding Council (UK)
UKAEA	United Kingdom Atomic Energy Authority
VITO	Vlaamse Instelling Voor Technologisch Onderzoek (Belgium)
VUB	Vrije Universiteit Brussel (Belgium)

Introduction

'Successful companies do not try to develop all their technology in-house'. They are aware of what is being developed both domestically and internationally, and are prepared to license technologies that can help them move rapidly into chosen markets. They also work collaboratively with suppliers and customers to develop technology and use academic institutions to tap into technological resources that they could neither attract nor justify employing in-house.

'Manufacturing into the 1990s' DTI\PA *Technology 1989*
(pages 40–41)

From the 1980s industry's acquisition of externally developed technology became a focus of public policy and of academic research. The sentiments in the UK's Department of Industry report quoted above regarding industry's need to capitalize on the knowledge resources in the science base were replicated in government reports from countries in the developed and the developing world. At the same time, academic studies in Europe and in the US were documenting the growing complexity and spatial diversity of the innovation process in which industry and academic links had become measurably stronger. Studies, like the one discussed in this book, were predicated on the assumption that the balance of innovation had in some ways shifted away from the firm as the focal point, into universities and national laboratories, which for the purposes of this book are described as the public sector science base (PSSB).

The book critically examines the phenomenon and the consequences of the increasing inter-dependence between industry, universities and national laboratories: the interface between private and public investment in technology. The increasing trend of use by industry of public sector resources is an example of *externalization of innovation*. This trend is arguably part of what Howells and Green (1988, 130) have described as externalization of the production process. The reasons behind the increase in industry and academic agreements have been explained by such factors as the increasingly rapid pace of technological change and the spiralling costs of R&D. Externalization is an attempt to spread the risk and sometimes the cost of innovation.

Moreover, the tendency in national and European policy towards funding collaborative programmes has intensified the trend towards collaboration as a form of externalization.

Externalization of innovation provides the potential for increasing returns, which according to Storper and Walker (1989, 55) is, 'the essence of technological change. The choice is between present and future techniques as industrialization moves forwards'. In other words external innovation links can be employed both as an industrial technique used to invest in the future, and as short term competitive strategy. It is part of the process by which firms in technologically dynamic industries maintain their competitive edge, and successful firms which are otherwise 'slow and clumsy in shifting out of old technologies'... 'move into new and superior ones' (Olleros and MacDonald 1988, 155).

This book is concerned with the interaction of geographical, social, organizational and political processes which combine to produce externalization of innovation. They are geographical because the scope of flows of information are related to the spatial construction of sectors and the location of expertise in the PSSB. They are social because as Sayer and Walker (1992, 115) have pointed out, 'innovation is fundamentally a social process built on collective knowledge and co-operative effort'. The process of organizational change has in some cases taken the form of fragmentation of R&D whereby contract research has boomed and small firms and small units have claimed a rising share of innovation (Whittington 1990, 184). This process has created complex patterns of interaction, breaking down the established patterns of universities and national laboratories primarily interacting with large firms (Howells 1986). At the same time political priorities have led to the increasing re-orientation of research in the PSSB towards the needs of industry. As a result, universities and national laboratories have become institutionalized into the innovation process *per se* (see David *et al* 1994; Lawton Smith 1997).

However, while externalization of innovation is a general trend it is not a universal process. This book investigates how patterns of linkage develop over time by identifying which factors regulate the extent to which firms chose to supplement their in-house innovation activities with expertise in the PSSB. It analyses the causes of inter-firm and inter-sector differences in three countries, the UK, France and Belgium, comparing the experiences of firms in a mature industry, the flow measurement industry, with those in an industry characterized by rapid technological change, the electronic components industry. The views of academics, scientists and engineers in the PSSB in the same countries involved in research activities on behalf of the two sectors are also represented. By examining the construction of links between industry and universities and national laboratories (the organization of innovation) together with where industrial innovation take place (the geography of innovation) the study contributes to understanding of the spatial dynamics of industrial innovation. The national, and to a lesser extent the regional, contexts provide the background to the story. The nationally variable sector-specific internal skill resources and intellectual capital of industry and the contribution of universities and national laboratories are shown to be important creative forces in shaping conditions under which interaction occurs.

Definitions

The analysis of which factors mediate in the process of externalization is developed from a line of approach suggested by Dicken and Thrift (1992, 280-7). They argued that in order to understand the organization of capitalist production there is a need to look at systems of firms. They suggest that production is organized primarily by business enterprises operating within extremely complex, dynamic networks of internalized and externalized transactional relationships of power and influence. These systems have also been called *filieres*, a filiere being 'a connecting filament among technologically related activities', (Truel 1980 quoted in Storper and Walker 1989, 133).

Innovation filiere

The underlying analytical concept used here to describe the system or chain of activities which links firms with universities and national laboratories is that of *innovation filiere*. Innovation filieres sometimes include industrial associations and user firms as participants. This narrow focus therefore omits other direct contributors to innovation including suppliers, subcontractors, and specialist R&D organizations. The definition of filiere which is used as a basis for defining an innovation filiere in this study was that suggested by Preston (1989, 6). A filiere is:

> a chain of activities which produces a given set of goods or services; it is the structure of a flow process linking raw materials with final consumption through a set of inter-related and interdependent stages. The coherence of a filiere is determined by a shared technical system of production and related know-how and skilled labour; by a shared industrial structure particularly of capital markets; a

common product or service market; and a common set of government relations. It is conceived as a 'structure in process' because at
every stage it is subject to pressures for change resulting from
changes in technology, market structure or government regulation.

The book examines how innovation filieres are constructed and continually reconstructed by a series of interdependent and countervailing
forces which include public policy, technological change, firms' innovation strategies, customer demand and the external competitive environment. The book pays particular attention to the role of government
in shaping innovation filieres in each sector in each country. This is
because other studies have found that successful industrial practices
around technology use arise to a large extent from deliberate state
action (Gertler 1997, 28). The impact of government regulation is
explored from two perspectives, those of 'regulation' in a specific sense
and institutional change.

The chain of activities which comprise innovation filieres comprises
three operational elements. These are networking, flexible collaboration and externalization (see Swyngedouw in Chapter 1 for a discussion of the relationship between technological change and these three
form of interaction).

Networks

Many types of networks of innovators exist (De Bresson and Amesse
1991). Here networks are considered to be primarily a source of information exchange. Not only can technical information be transferred
during informal interaction, but other sorts of useful information can
be communicated. They include information about what kind of
research is being done by whom and where and what kind of approach
is necessary to gain access to it. Information gathered and disseminated
through reciprocal exchanges denotes a particular kind of leverage
which is operated by individuals and collectively by firms when they
access and contribute to networks. The term is not used in the sense of
'an organizational method of managing technological innovation'
(Albertini and Butler 1997, 11). This use has more in common with our
definition of collaboration.

Collaboration as a means of innovation is derived from organizational
strategies and contributes to and builds on networks. Its basic characteristic is the sharing of expertise through working in association with
other organizations. Formal relationships, such as those under national
and EU innovation support programmes, frequently arise from a

shared set of intentions identified during the course of informal net-working.

Externalization as defined here differs from collaboration in two ways. First, externalization involves the utilization of existing information rather than the generation of new knowledge which is an outcome of collaboration. It requires that the PSSB accept in principle as well as in practice that they are there to solve industrial problems. Second, the contractual basis of the relationship defines the deliverables in terms of tasks and outcomes within a limited time-frame. Thus there is an in-built rigidity because externalization is less likely to be 'curiosity-driven' than collaboration.

The difference between the first two concepts and externalization is that the former are ideas of renewal and flow of information whereas the latter relates to the exploitation of stocks of information. It is argued that the operation of the first two overcome the frictional powers of time (the rate of technological change) and space (location of technology and its appropriability) while as the latter is driven by a lack of internal resources and is therefore most price sensitive, it is more likely to be within domestic space (see Chapter 1 for a more detailed discussion of this point).

The three forms of commitment create what Swyngedouw (1996, 12) has elsewhere called changing 'socio-spatial relationships' in which new arrangements embody a set of power relations and norms of co-operation. These arrangements result from different kinds of power and influence (political, economic, social and organizational) operating at particular moments in time. They influence the location of external-ized activities and contribute to changes in the composition of the skill base in both industry and the PSSB.

Regulation

The term 'regulation' is used here in the specific sense of the kind of governmental intervention which affects firms' innovation strategies and indirectly their propensity to seek additional technical resources from the PSSB. It encompasses more than the explicitly interventionist stance of 'technology policy' which is defined in Chapter 2, but is a more limited use of the term which economists use to denote interven-tion by government in economic behaviour through laws, entry regu-lation, the setting up of regulatory agencies and other instruments such as competition policy, all of which are outside the scope of this study (see Hay and Morris 1991, 622, on the types of intervention and the rationales for regulatory intervention).

To help identify the impact of regulation on the construction of innovation filieres, four questions provided by Yarrow (1996, 316) are used. These are:

- Who regulates?
- What is being regulated?
- What instruments of regulation are used?
- How is regulation enforced?

The question of what is being regulated here refers to the structure of innovation filieres, particularly the degrees of cohesion and mutuality of objectives in the relationships between different organizations and their 'efficiency' as judged by the criterion of their contribution to innovation.

'Instruments of regulation' are those formal mechanisms which determine the possibilities for the direction of scientific and engineering research in the PSSB and those which under which the terms of trade with industry are determined. They include the rules under which scientific and engineering research in the PSSB is conducted and owned, as well as a range of statutory instruments such as standards and environmental regulations.

Formal instruments and institutional change are only two components of control or regulation of the interface between industry and the science base. Other practices and structures can be important. In another context of regulation, Hancher and Moran (1989, 291) have pointed out:

> that in any regulatory arena there exists a diversity of organizations which interact through a variety of networks of varying intensity of formality. ... these linkages may be articulated in terms of formal, binding legal rules, standard operating procedures, or indeed the forms of more conventions. The range of those involved in the regulatory process will not only vary from issue to issue but will also be deeply influenced by convention or constitutional practice, or by prior patterns of regulatory practice. Inter-organizational linkages are the subject of a large, existing literature on network theory which demonstrates the significance of policy communities and networks within which 'elite coalitions' allocate issues to particular arenas, manage the policy agenda, and allocate the range of participants allowed into decision making.

This view of how regulation is effected can be equally applied to the analysis of how innovation filieres are regulated. Of particular interest are the outcomes of the power exerted through the representation of different groups in the policy making process and which decide allocation of resources to technological fields. Regulatory systems are enforced not only by rules of entry and conduct but also by the expectations of industry and scientists and engineers in the PSSB of how the system operates.

Institutional change

Institutional change here refers to the changing structure and function of universities and national laboratories and hence their roles within innovation filieres. The changes in the supply-side of technology transfer is a central theme of the book.

The study

This book returns to an earlier tradition of sector studies. Since the 1980s geographers' attention has moved away from looking at the spatial organization of sectors to focusing on the internal dynamics of innovative regions. Often this has been considered within the context of the internationalization of production and technology (see for example Dicken, Forsgren and Malmberg 1994). Here the interest is in the spatial construction of innovation activities within and across countries and the extent to which innovation filieres coalesce in particular places. Evidence from the US shows that almost all industries are to some degree localized. However, in many industries, the degree of localization is slight, particularly in mature industries (Ellison and Glaeser 1997, 889). Moreover, these authors found that geographic concentration resulting from choice made about the location of production does not imply the existence of intellectual or technological spillovers which are assumed to be implicated in the growth of high-tech industry in such places as Cambridge (Segal Quince 1985, Keeble *et al.* 1997, Garnsey and Lawton Smith 1997).

The empirical research recorded here provides both an important record of recent trends and of the broader context in which interaction takes place. The comparisons of conditions under which links between the science base and industry develop are used to highlight differences in (i) national and sectoral patterns of innovation in industrial organizations, and (ii) the ways in which resources in the PSSB are used by industry. The flow of information between industry and national laboratories and universities is usually seen in terms of transfer *out* of universities. In reality, what is increasingly happening is that ideas and problems of industry are also transferred *into* the science base. This has important implications for the control and direction of scientific research.

The sectors

The two manufacturing industries are important for two reasons. The first is their importance in relation to other industries. Although the value of the production of the flow measurement industry is small compared to other sectors, it is nevertheless a pivotal industry. It is one of the most important process measurement variables, being used for monitoring, control, fiscal and legal purpose in a wide range of industries and utilities (Sanderson 1989). Its products are in constant demand, both in industry (as all industrial sectors measure flows of gases, liquids and solids) and in the domestic water and gas utilities. It is an industry faced by opposing pressures. On the one hand, the recession of the 1980s saw a withdrawal by many users of instrumentation, for example petrochemical companies from the design process and an increased demand for simpler and cheaper instruments. On the other, there were demands for new flow (and level and pressure) instruments and techniques to be developed to meet the requirements of modern process plants. This involves much more widespread use of sophisticated electronic technology. A further technological shift is the development of 'intelligent' pipes, replacing the need for meters.

The electronic component industry is strategically important in a raft of industries. The success of all applications industries – consumer, IT, defence, control and measurement, telecommunications and now automotive depend critically on access to and innovative use of component technology (NEDC 1991, 2). Indeed just under one-third of inputs into electronics industries are electronic components (NEDC 1991, 6).

The second reason is that they represent a contrast between an industry which has relatively undeveloped linkages with universities and national laboratories (flow measurement) and one with well established and varied forms of interaction (electronic components).

Methodology

The study tested five sets of assumptions. These were that there would be:

(i)   Increasing research interaction with universities and government laboratories as a response to firms' recognition of the value of tapping the science base as a competitive strategy.

(ii) Country/regional differences arising from past regulatory practice.

(iii) Significant sector differences in organization of linkages.
(iv) Differences at the firm level.
(v) Regional differences in access to technology.

These were tested by analysis of interviews based on questionnaires with:

- samples of firms from each country selected to represent characteristics of size and parent ownership, and different technologies within the two industries. The total sample size in flow measurement was 25 and in electronic components 23 firms.
- samples of universities and national laboratories whose research interests matched those of the firms. The samples consisted of 12 departments in flow measurement and 14 in electronic components.
- representatives from government organizations (national, regional and EU) with direct interest in or responsibility for promoting innovation in industry and/or fostering links between industry and PSSB.
- trade associations.

In both industry samples, the majority of the interviews were with senior technical staff (scientists and engineers) and senior managers. The exceptions were in the flow measurement sample in Belgium where they included sales staff, and the managing director of a Belgian chemicals company. Interviews in academic and national laboratory samples were with scientists and engineers except in two national laboratories specializing in electronics, one in Belgium, the other in France, where they were with senior personnel and marketing managers.

The industrial samples were chosen to reflect the range of size and ownership and product areas. They also reflected the structure of the filieres in each sector. Thus in the flow measurement industry, user firms such as petrochemicals companies were included as traditionally they had been closely involved with the design of meters. Although the demands of users shape innovation strategies in the electronic components industry, the closer relationship is at the technological interface between component manufacturers and scientists and engineers in universities and national laboratories than with individuals in user firms. Only one user firm was included in the electronic components sample. This was the research laboratory of a UK aerospace firm. It was included because it had been involved in the design and production of electronic components and still had a close interest in the design of components used in the defence industry. The small samples provide examples of innovation strategies and are indicative of more general patterns of institutional and structural arrangements which account for geographical change.

Organization of the book

Chapter 1, written by Erik Swyngedouw, provides the theoretical perspective on the territories of innovation expanding on the theme of collectivization of innovation. He argues that the success or failure of regional or national economies is to a large extent dependent on their capacity to generate continuous technological and organizational innovation in the context of global flexible competition. Moreover, the speed of technological change and the uncertain success of technological strategies favour processes of networking, flexible collaboration and externalization. On the one hand success depends on the collective structure of the economy and on the other the profit rate and growth of each individual firm. Erik highlights the characteristics of territories which shape collective and individual innovative capacity.

This perspective is complementary to the theory of regional adjustment (Clark *et al.* 1986, 2) which considers how economies work in time and space. The theory of adjustment is an attempt both to explain how economic uncertainty is absorbed by the spatial organization of production and to show the implications of such organization for the structure and dynamics of regional economies. A key element of this theory is that the arrangement of uncertainty has dramatically changed with equally dramatic shifts in the nature of industrial organization.

In this book we show that a principal source of uncertainty within innovation filiere is the allocation of resources to finance innovation by both firms and by regional and national governments and, in Europe, the EU. Spending patterns are of major significance in the supply of and demand for technology. This basic fact can be overlooked when other factors such as spatial proximity and social factors are the prime focus of investigation.

Chapter 2 examines the regulatory context to interaction. This summarizes recent EU, national and, in France and Belgium, regional innovation policies. This provides the framework for examining the impact of these policies on both the geography and efficiency of innovation filiere. It begins by discussing the reasons for political intervention in the process of innovation before moving on to outlining EU policy and programmes in flow measurement and electronics. The chapter then introduces differences in national investments strategies and details the main features of the three national innovation systems including the organization of the science base, support for innovation in industry and mechanisms to support the interface in both sectors in each country.

Chapter 3 provides an overview of trends in industry and academic links and reviews the changes in the PSSB in each country, focusing on institutional change comparing the more fragmented system in the UK with the more centralized systems in France and Belgium. In the latter two countries national/regional authorities have acted to produce examples of spatial integration of industry and public sector research, particularly through the activities of two national electronics laboratories. These are the Laboratoire d'Electronique, de Technologie et d'Instrumentation (LETI) in France and the Inter-university Microelectronics Centre (IMEC) in Belgium. In both cases historically embedded technological trends are being underwritten by regulatory action as spatial clusterings of industry and PSSB links have been fostered and legitimated by regulatory frameworks operating at the regional level in parallel with national, EU and other international programmes designed to encourage interaction. Thus the capacity, quality and geographical focus of innovating networks is influenced by the multiplicity of interacting and countervailing local and national economic and political vested interests in the face of economic and technological uncertainty (see Swyngedouw 1994).

Chapter 4 focuses on factors influencing industrial innovation activity. The conditions which other studies have found to inhibit or encourage individuals in firms or firms as a result of articulated innovation strategies, to have informal and formal links with the PSSB are discussed in the first part of the chapter. The second part describes the specific characteristics of each sector in order to identify differences between the two sectors in the modes and degrees of innovativeness (see Dosi 1988, 222) and from those, the general propensity their firms to interact with the PSSB (c.f. Faulkner and Senker 1995). Collectively interdependent features include the relationship between ownership, innovation strategies and market structure, and how the collective might of sectors can lever advantage within the political system through membership of 'elite coalitions'. If technological change is about the fusion and diffusion of knowledge then the reason why some sectors are better able to cope with advances and discontinuities in technological advances can be seen in terms of managing the problems of access to information. This includes over-coming a number of appropriation issues such as managing the flows of information needed to employ the gains from convergences in technology, such as in telecoms and computers, and developments in new generic technologies, participation in internationalization of research driven both by scientific advance and by funding agencies such as those of the EU.

Against that background, Chapters 5 and 7 each contribute to identi-fying how innovation filieres are constructed by recording the reality of the experiences faced by firms using evidence from a series of case studies. They do this by examining the overall geography of linkages, sources of external funding and the way that sectors are co-ordinated and represented in decision making. Chapter 5 discusses the flow mea-surement industry and Chapter 7 the electronic components industry. The evidence is used to examine how the composition and dynamics of innovation filieres arise both from the actions of firms, individually and collectively as a sector, as they respond to the competitive envi-ronment. It shows how an individual firm's innovation strategy can reflect both the characteristics of the industry in a particular country as well as its technological position within that industry (Bye and Chanaron 1995, 63). It is argued that over time, new categories of groups (firms, elites) assume more powerful positions of influence as the focus and location of research activities shifts within dominant groups as others withdraw from particular activities.

Chapters 6 and 8 reflect the views from groups working with indus-try from the respective science bases. In both cases extensive networks exist but while externalization is the more common means of inter-action in the flow measurement industry, collaboration is the norm in the electronic components industry. Chapter 9 reassembles these con-stituent parts and summarizes the evidence on which factors regulate the structure of innovation filieres. It highlights the reasons behind the pattern of more spatially extensive set of linkages in the electronic components industry than in the flow measurement industry.

To conclude, the study raises a number of issues about the ways in which innovation filieres operate. The presumption that efficient inno-vation systems in which industry and PSSB links will be the medium through which industrial competitiveness is achieved tends to blur the distinction between complementary activity and that which is a substi-tute for industrial research. This book argues that the commercializa-tion of public sector science requires that ideas and problems of industry are first transferred *into* the PSSB before there can be reverse flows of information out of universities and national laboratories. Commercialization of PSSB resources is not without cost. The study was as much about the tensions and uncertainty it causes rather than overcomes as with the forms through which interaction takes place.

Part One
Background

1

Territories of Innovation: Innovation as a Collective Process and the Globalization of Competition

Erik Swyngedouw

1. Homing in and spacing out: the new competitive environment

The forms of production organization, of production process logistics, of defining technological and organizational trajectories, have undergone rather dramatic changes over the past two decades or so. In addition, competitive mechanisms and systems of technological innovation have also changed in innumerable ways and are expected to continue to do so in the course of the foreseeable future. The consumer commodities and equipment producing industries, which underpinned post-war economic growth in most advanced capitalist countries – automobiles, chemicals, heavy engineering – are no longer considered to be the linchpin of national economic recovery and modernization. They have been replaced by advanced knowledge-intensive industries and services; activities whose success is predicated upon continuous technological innovation and organizational adjustment.

Indeed, there is a widespread consensus in the literature now that the post-war vertically integrated, self-centred, technologically 'locked', standardized mass-production based industrial giants are gradually losing their predominance in favour of vertically disintegrated, intensely interacting and rapidly changing (in terms of organizational and technological practices) production complexes. The latter's organizational unity is assured through a myriad of external relationships and linkages (expanding and deepening sub-contracting relationships, joint ventures, strategic alliances, out-sourcing) (Storper, 1997; Cooke and Morgan, 1998). Of course, large global companies have not disappeared at all and will not become extinct either, but have turned into global lean, mean, money making machines, leaving 'local' companies with the nitty-gritty of everyday production under the watchful eye of these globally competing industrial or financial power houses (Harrison, 1997). 'Glocal' strategies, that is strategies of maintaining strategic world-wide control, are achieved through and combined with strategic insertion into highly innovative 'local' territorial milieux. Through external linkages and transactional networks, global companies plug into the innovative dynamics of particular places, regions and countries in order to assure continuous innovation and, hence, the maintenance of a globally competitive position (Swyngedouw, 1997a,b).

The advent of global competition, in which regional or national economies are nothing but strategic places in a continuously shifting and expanding multi-dimensional chess game, has affected the rules of the competitive process in important ways. Global competition for a global market place has intensified the struggle of individual companies (both large and small) to survive (Scott 1998). It is no longer the competitor down the road which may threaten a firm's sustainability, nor even the position of the French, Italian or German players (although of course they are still important), but it is the Americans, the Japanese, the South Koreans and pretty soon perhaps Australian, Brazilian or Filipino producers.

In the context of the struggle for global markets, cost minimization and price competition is the assumed rather than the agreed upon strategy. Quality and substance of the commodity have become crucial elements in competitive strategies. Producing new technologies and developing new organizational principles determine long-term success in terms of increased quality and innovation of products and, thus, economic growth. Product life cycles are increasingly reduced; products rarely reach a stage of maturity as they become obsolete or are out-competed before they can become fully standardized. Continuous technological change is the hallmark of today's competitive struggle and determines a company's and a country's economic success. The new competitive environment requires companies that are simultaneously highly specialized while not being locked into a given technological paradigm of system. Technological lock-in is a suicidal course of action. This holds true both for highly innovative sectors in which the commercial prospects of new technologies are highly uncertain, such as in electronic components, as well as in a more 'mature' – but now rapidly de-maturing – industry, such as the flow measuring industry.

This transformation, which has barely begun, puts the finger on the key problem this book attempts to address. The success or failure of regional or national economies is to a large extent dependent on its capacity to generate continuous technological and organizational innovation in a context of global flexible competition in which a) each place attempts to achieve the same capabilities; and b) each individual firm increasingly depends on other companies (and institutions) to generate a sufficiently high innovation pace to outperform its competitors.

In short, the very speed of technological change and innovation and the uncertain success of technological trajectories favours processes of externalization, networking and flexible collaboration rather than integration and internal control. Thus change depends on the collective structure of the economy on the one hand, and on the other hand, the overall success of an economy is dependent on the profit rate and growth capability of each individual firm.

It is exactly this conundrum I wish to focus on in this chapter. While global market Stalinism has become the latter-day gospel of international organizations, and governments around the world promote the interest of private and de-regulated companies, the innovation process is rapidly becoming a more collective and socialized process. Co-operation between private companies, between firms and local institutions (whether research, educational, regulatory or service based) and between firms and the local or regional collective social and physical infrastructure are central to innovative success (Amin and Thrift 1994; Storper and Salais 1997). At the same time, the global competitive environment demands a strong monopolization of innovative capacity and disciplines those who are unable to keep competitors at bay. In the first part of this chapter, we will focus on the process of externalization and attempt to theorize the territorial or spatial foundation of this process. In the second half, we will attempt to put some flesh on the notion of externalization, with particular emphasis on the externalization of the innovation process and the related problem of technology diffusion. This analysis will permit:

(i)   an understanding of how economic success depends crucially on the finely tuned balance between collaboration and collective action on the one hand and an individual company's strategic posture on the other;

(ii)   the identification of the institutional and structural arrangements which account for geographically divergent or uneven dynamics of technological innovation and diffusion.

2.   Towards 'glocal' competition: the dynamics of continuous technological innovation

2.1.   Technological competition in contemporary capitalism

Under capitalism, competition is the key driving force behind technological change and innovation. This struggle between individual units of capital can basically take one (or a combination) of three possible routes.

1   Direct price competition within a given set of technologies constitutes route one. This form of competition – Walker (1988a,b) refers to it as 'weak competition' – is usually associated with labour cost reduction, either directly (reducing wages through pressure on local labour or through the search for production locations characterized by a lower cost of labour) or indirectly (streamlining of the production process). From this perspective, the competition between individual capitals is mediated by the competition between capital and labour for the produced (added) value.

2   Another route is price competition in a context of changing technology. This strategy is, in fact, a variant of the first one in the sense that technological change may result in either a reduction of direct labour costs (through capital/labour substitution and productivity increases) or a reduction of indirect labour costs (i.e. through the use of cheaper inputs or production technologies).

3   The third route is competition based on product innovation (what Walker (1988a) refers to as 'strong competition'). This strategy is rather different in the sense that it attempts to make the competitor's product obsolete and, hence, generates a product-monopoly position for the innovator and poses, consequently, an existential threat to the non-innovator. In addition, new products create a new demand and, thus, a new market, which may have a negative effect on market demand for already existing products. However, this is a high risk strategy given i) the a priori uncertainty of the success of a new technology, and ii) that its success depends on rapid diffusion and standardization. The latter prompts innovators to give up their monopoly position in order to assure market penetration (cf. P-2000 versus VHS in video-technology).

The essence of successful competition is, therefore, continuously to try to create a monopoly position. In other words, successful competition is based on exploiting superior conditions of production to which the competitor has no, reduced or only time-lagged, access. A perfectly competitive market is a myth. If all firms can pay similar wages, use similar technologies and organizational systems, or produce similar products, profit would not be possible (Storper and Walker 1989). Only the continuous perversion of free market rules enables profit and, hence, growth. Technological change and innovation is, consequently, nothing other than a strategy to have 'something' the other does not have access to; it is conditioned by the production of continuous monopolies and continuous dizequilibriums. Perfect information, perfect knowledge and the unrestricted diffusion of know-how – one of the assumptions of a free-market based system – annihilates competition and therefore makes it impossible for one company to 'outcompete' the other. An economy in equilibrium conditions would, *de facto*, be an economy without profits and in crisis.

In addition, know-how and information are not 'free' goods. On the contrary, information and know-how (technology) are crucial, highly valued and costly assets. It is exactly this cost of know-how that permits monopolization. Making information freely accessible leads to the so-called 'free-rider' problem. The free dissemination of newly developed know-how would permit others to cash in on the success of someone else's work (and, thus, capital), thereby preventing the latter from achieving a monopoly position.

Nevertheless, the market does have an important function in undermining and breaking down existing monopoly positions. While innovation permits temporary monopoly conditions, market forces tend to undermine enduring monopolies (with a few exceptions such as natural monopolies). Maintaining profitability in a dynamic environment necessitates undermining the other's monopoly position through the introduction of new or better technological-organizational systems. A fundamental process in a capitalist economy is, consequently, the continuous production and subsequent destruction of monopoly positions.

2.2. The changing rules: from price to product competition or from scale to scope

The preferred competitive strategy, however, changes over time as competitive conditions and environments change. Price competition, although still important, is increasingly losing is predominant position. It is used as an integral part of the second and, particularly, of the third route to competitive advantage. First, lowering direct labour costs is problematic under conditions in which a) wages are regulated and can only vary marginally and – more importantly – b) there is virtually no opportunity to monopolize low wage advantages for a more or less sustained period of time given the increased locational mobility of investments. Second, indirect labour cost reductions through technological-organizational change are relatively quickly imitated and, consequently, result in very short-term monopoly positions.

Indeed, the possibilities of achieving and, more importantly, of maintaining a monopoly position have dramatically changed in recent years as the scales at which companies operate have changed. Maintaining a monopoly condition is scale- and scope-dependent. The former refers to the geographical scale over which a monopoly position needs to be maintained. The globalization of the economic process, the increasing scale and size of individual capitals and the 'footlooseness' of investments has meant that many more players are trying to undermine existing monopoly positions than was previously the case. This globalization, consequently, increases the speed by which new technologies need to be developed and introduced.

The scope-effect refers to the range of activities (technologies, products) in which a monopoly position needs to be maintained. The more specialized, upmarket and specific a technology or product is, the easier it is to maintain or consolidate a monopoly position. Once 'locked' into a particular technological route, it is extremely difficult for others to imitate (given a variety of barriers of entry). Moreover, it is equally difficult for the innovating firm to change its course of action without major financial sacrifice.

However, in the context of global and rapid technological change, the process of monopolizing new technologies becomes more costly and risky as a) the products become more sophisticated and require a great variety of know-how and skills and b) the prospects of commercial advantage (acquiring a monopoly position) are more risky (as someone else may be there – or come up with a better product – before you). Moreover, locking into a given technological and organizational trajectory (and all its associated costs and rigidities) may inhibit, stall or slow-down the speedy introduction of a 'new' set of technologies.

The globalization of economic life has spearheaded the increasing importance of the third competitive route at the expense of the other two. Vertical integration traditionally favoured price competition as a means to achieve monopoly positions. However, this process simulta-neously reduces the speed at which change can be introduced, increases the cost associated with developing and introducing new technologies and production processes, increases the cost of failure, demands long incubation times, and tends to lock firms into a pre-existing technological paradigm or trajectory. Technologically dynamic industries, in contrast, tend to be organized in dense organizational networks, characterized by complex, yet flexible and continuously shifting, external relations with other firms, agencies and/or institu-tions. This organizational structure permits continuous learning and innovation in a context of continuous restructuring and complex, but shifting, divisions of labour. Such networks permit individual firms to externalize, at least partially, the costs associated with maintaining such dense relational complexes. Continuous innovation demands organizational flexibility and favours processes of disintegration and externalization (Scott 1988). However, such an externalized transac-tional system does not exclude patterns of domination and subordina-tion between global and local firms; in fact it presupposes such relationships.

The combination of intensified global competition and a rapidly changing technological environment prompts the above-described pattern of externalization as a cost-minimization strategy under condi-tions of continuous uncertainty (Camagni 1990; Storper 1997). It has become increasingly clear that this process of externalization is dis-tance- and space-sensitive. First, the transactional nature of the exter-nal relations (speed, cost and quality of the external organization) is of paramount importance in setting the dynamics of the externalized set of relations (Scott 1988) and in determining the aggregate level of cost minimization obtained by the transacting firms. The distance-sensitiv-ity transcends pure agglomeration economies obtained through dimin-ishing material transaction times, but also includes processes of continuous bargaining, negotiation, reinterpretation and re-definition of the practices of inter-agent interaction as rules, regulation, norms, codes, and know-how are subject to continuous change (Storper 1997). Second, the space-sensitivity, or territorial sensitivity, lies exactly in the geographical character of the physical, institutional and socio-cultural carriers of the transactional flows. Indeed, under conditions in which external relations have a growing importance in underpinning the dynamics of change, they equally escape the control of the individ-ual firm as those relations become 'embedded' in a system of negoti-ated and essentially collective relationships. In other words, the capacity and quality of the innovating networks will be defined by and depend on, among others:

- the degree, level, and quantity and quality of access to other actors
- the characteristics of regulation of capital markets, labour markets, and capital/labour relationships
- the codes and practices of supplier-buyer relationships
- the nature, interpretation and organization of formal (contractual) and informal agreements, the agreed-upon levels of quality of work, the influence of third party institutions such as research labs
- the nature of the local, regional or national state
- the cultural and ideological practices and the scale at which these practices operate.

It is these spatial characteristics, which shape the collective and individual 'innovative' capacity associated with externalized networks of relationships. In other words, the 'external milieu' which carries the transactional flows will set the conditions of innovative capacity within that territory. Those elements did not play the same paramount role under conditions of vertically integrated organizational structures under which the above-mentioned characteristics were often internalized; that is subject to privatized control under a unified command centre. For externalized processes, the characteristics are produced through collective control and command (Swyngedouw 1992).

In short, the increasing dependence on external relations seems a precondition (at least at this moment) for continuous innovation and organizational flexibility. The price (and risk) to be paid for the improved technological capability is the simultaneous reduction in command and control of individual firms over technological trajectories and increased dependence on the external transactional structure. The latter prevents full control and exclusive access and demands a certain degree of negotiation, bargaining and collaboration.

We are faced here with an apparent paradox. It is exactly the process of globalization and the ensuing increasing speed of product innovation to maintain a sustainable competitive position, which resulted in the process of externalization and the generation of territorial network complexes. The capacity of the latter, in turn, to create an innovative and technologically dynamic transactional configuration does not depend on the individual strategy of individual companies but rather on the quality and nature of the inter-agent relationships as defined above. Since the latter are highly space sensitive and geographically differentiated, territorial innovation dynamics vary widely. However, the long-term economic success of regional and national economies in the global market place increasingly depends on their success in generating particular territorial configurations; hence the divergent growth dynamics of France, Belgium or the UK and, within those, of the regions of Rhône-Alpes, Flanders or South Wales.

For the individual entrepreneur operating within a territorial configuration, there is always a choice of competitive strategies, depending on the nature of the external 'milieu'. However, such choice embodies a series of possible trade-offs. For example, if a firm can maintain profitability through labour-cost reduction, there will be less of an incentive to opt for the third route of continuous innovation. This is exactly at the heart of the UK's current problems of competitiveness. Opting for a low wage strategy through renegotiating capital-labour relationships (route one, cf. Rolls Royce, for example), reduces the short-term necessity to engage in a route three type of strategy. This preference, which is set by institutional codes and rules – such as wage settlements procedures – reduces innovative practices and – in the medium term – negatively affects the competitive position in the global market place, in particular *vis-à-vis* those who take another route. For example in Germany, where route one is closed off through strict wage regulation, increased product innovation strategies are a necessary competitive strategy, resulting in a better medium-term performance in the global market place.

Moreover, the importance of such configurations for growth should not only be assessed in terms of endogenous (or local) firms, but the externalization practices of global firms (which in the end have to insert themselves into such networks) in particular territorial configurations will play a key role in setting the long-term development trajectory of these territories in the international division of labour. This 'glocalization'-strategy in which global firms (have to) embed themselves in particular territorial transactional networks is influenced by the differences in innovative capacity of those networks (Swyngedouw 1997a). IBM will not insert itself in a transactional territorial structure that exhibits a fairly weak innovative capacity, while – vice versa – the capacity of the local milieu will be enhanced by IBM's decision to insert itself within a particular network, resulting in cumulative growth in one place at the expense of another. In short, competitive strategies of firms are increasingly based on homing in on highly innovating territorial environments in order to be empowered to space out globally (Swyngedouw, 1998).

In short, the success of a regional or national economy depends on its ability to maximize the third route of competitive strategy and, hence, on the development and quality of such flexible, externalized organizational structures. Local conditions of innovation play an increasingly important role in maintaining global competitiveness (compared with 'traditional' internalized innovation practices). However, creating these conditions is not so obvious for a number of reasons. It is these that we shall turn to next.

2.3. The contradictions of permanent innovation

One of the tensions, which are generated through this growing role of the transactional environment, is that the externalization process leads to a situation in which the innovative capacity is collectively generated. In other words, innovative potential is often determined by the collective and, hence, socialized interaction between individual agents. The collective effect produces better results than the sum of the individuals (that is exactly why externalization seems to be preferred compared with internalization). The success of the individual company in the market place, however, depends on the success with which it can privately appropriate and monopolize (that is to exclude others from gaining access to) the new collectively generated know-how. Put plainly, the success of a company depends on the success it has in excluding others from accessing a particular, yet collectively generated, know-how.

In sum, a fine line, a trade-off, between co-operation and competition will determine the success of the overall innovative capacity. The mechanism by which this tension is contained seems to me to be crucial in assessing innovative capacity. In the literature, a clearer picture seems to emerge gradually in terms of the variety of mechanisms through which this tension is contained and the relative success of the ensuing articulation between the private and the collective. Exclusive competition within networks seems to have a devastating effect on medium-term innovative capacity and, consequently, on maintaining or consolidating a competitive position in the global market place, while intra-network co-operation and collaboration (within the territorial configuration) maximizes innovative capacity. The latter strategy diminishes the relative monopoly position of each individual firm within the transactional network, but improves the monopoly position *vis-à-vis* other (Japanese, American, and so on) transactional configurations. In fact, that is exactly what underlies the success of the Japanese or, for that matter, of much publicized stories such as SWATCH in Switzerland or BENETTON in Italy: collaborate at home and compete abroad or ... home in to space out.

The success of competitive strategies along the third route (product innovation), therefore, is no longer conditioned by inter-firm competition *per se* and its mediation through the struggle between capital and labour. Inter-territorial struggle – territorial here defined as an 'embedded' collectively produced transactional network, a 'milieu' – mediates in the success of individual firm in keeping afloat in the global competitive environment. In the end, society does seem to be something more than just a sum of individuals pursuing their private interests!

Having established the role and importance of the system of external transactional relations and their support structures, we now turn to a discussion of these externalities themselves. What are they? Where do they come from? What do they mean? This is particularly relevant in the light of the symptomatic absence of a theoretically rigorous discussion of externalities in the economic or geographical literature.

3. External relations, inter-spatial competition and the private/social dilemma

3.1. Externalized relations and forms of territorial configuration

The transactional characteristics of external relations are structured through a form of 'territorial organization', which comprizes:

(i) the 'general conditions' which enable individual firms to transact, to engage with their external environment. This refers to the physical infrastructure (roads, (tele-)communication infrastructure and the like) as well as to the organizational-institutional structures which carry the circulation process (training and education centres, research institutions, and the like) (Lapple 1976);

(ii) the 'institutional' or 'regulatory' infrastructure. This refers to the territorially developed norms and rules, which regulate the interaction between individual agents or social groups and is embodied in formal (legislation) or informal codes. We can label this 'institutional' infrastructure as the form of 'territorial regulation' (or micro-regulation as Getimis and Kafkalas 1989 11 call it). This includes, among others, the nature and organization of the capital/labour relationships, and the role of the state and other institutional carriers such as development agencies, technology centres, Chambers of Commerce, etc;

(iii)   the characteristics of the labour force and, in particular, the set of qualifications and skills and their formation;
(iv)   the characteristics of the units of private capital with their specific internal and external technological-organizational structure.

Obviously, these elements of 'territorial organization' are not independent from one another, but are in fact constructed through interaction between individual and social groups. As such territorial organization is a collective result, the outcome of social interaction. Classical economists usually identify these collective effects associated with territorial organization as 'externalities'. In fact, these 'externalities' are nothing other than the expression of the socialized nature of capitalist production and consumption (Gordon 1989). This means that in order to produce and to consume, there is a need for a series of collective material and immaterial infrastructures.

J. C. Perrin, for example, defines externalities as:

> The global collective advantages [or disadvantages, I would add] which result from the economic combination and spatial convergence of diversified and complementary productive equipment, superior business and collective services, and industrial and administrative structures which assure communication and concentration. This advantage derives from an organizational process, which is different from that of the market. They (externalities) produce a structural and indivisible commodity which confers capacities upon a macroeconomic ensemble which cannot be attained without them (Perrin 1974, 36).

This superior collective commodity derives exactly from what we have labelled as 'territorial organization' defined simultaneously by a territory's (whatever its scale) economic, socio-institutional and spatial content. This 'territorial organization' manifests itself through quantitative effects (agglomeration, scale, multiplier, infrastructure effects) but also through qualitative effects (diversified activities, bundles of information centres, networks and flows, decisions centres, market organization and characteristics, and so on) (Perrat 1987), each of them supporting the flow of production and consumption. Such organizational pattern results, therefore, in a creative ensemble, external to the individual firm but internal to the processes of economic growth and expansion and enhances the productive capacity of the socio-economic space in which the ensemble is rooted. The classical notion of 'externalities' is, consequently, extremely ambiguous. These effects are perhaps external to individual actors, but are fundamentally internal to the process of innovation and economic expansion.

3.2.   Co-operation and collective effects versus private appropriation

External effects, as defined above, derive from the necessary socialized character of production and consumption in a context of private appropriation of the benefits through the market. Integrating external effects (through vertical integration) privatizes the benefits of the effect but simultaneously privatizes the cost. Processes of externalization socialize the cost partially, but permit others to cash in on the generated external effect. The dialectics of internalization/externalization of these effects, exemplified by the dialectics of vertical integration and disintegration (see Harvey 1982), express the contradictory nature of the dual 'social-private' character of capitalist production and consumption. Internalization invariably increases overhead cost (fixed capital embodied in infrastructure, logistics and logistics organization, planning and management), increases the initial capital outlay, slows down the circulation time of individual capitals, and may threaten the overall rate of profit of the individual agent. It equally assures the private appropriation of the internalized 'externality' effects. The process of externalization or socialization of these effects may result in a falling turn-over time of individual capitals and decrease the circulation time of capital in general, in decreasing risks of forced devaluation of capital, in increasing organizational and technological flexibility and so forth. Nevertheless, externalization opens up the ability for other (competing) agents to cash in on these socialized effects.

The dialectic of internalization/externalization of 'external effects' also entails the possibility of the continuous production of what Catin (1985) defines as 'latent externalities' or not-exploited externalities. Transactional configurations always contain a number of potential externalities. For example, one can decide to transact with a research centre. New transactional or externalized relationships permit the private appropriation of the capital embodied in the development, renewal and maintenance invested by other (private or public) actors, including that of past generations. It is clear that this process involves more than pure investment in the 'built environment' or 'physical infrastructure'. It may and does include, for example, training and education of the work force, research and development laboratories, networking agencies, institutions enabling or supporting material and

immaterial transactions, and so on (Peck 1996). Clearly, the production of territorial configuration can be the unintended (socialized)
outcome of individual capital's decision-making or be actively con-
strued through collective intervention (the state, growth coalitions,
business organizations and the like).

The territorial character of the transactional structure and the
'embeddedness' of external relations have as a particular character that
– contrary to other forms of productivity/innovation enhancing factors
– they are generally not commodified. In other words, external relations cannot be bought and sold on the market (and thus spatially
transferred) in the same way as information, know-how and technology can; they are – in Storper's words – 'untraded interdependencies' or
'relational assets' (Storper 1997). They are indivisible commodities,
whose effects can only be appropriated through insertion in the relational structure of the 'milieu'. The innovative dynamics of a set of
transactional relationships is 'located' and cannot be transferred to
other places and times. Growing externalization, therefore, assumes
the development of forms of co-operation and an expanding division
of labour. It is exactly this combined effect of the experiences and
qualities of co-operating, but specialized labourers, which permits the
continuous improvement of the means of production (which, through
linkage effects, ripples throughout the economy), reduces waste,
increases the time of utilization of the means of production, reduces
the turnover-time of capital (Marx 1977, 80), and is often a source of
innovations and inventions (Marx 1977, 104). This learning-by-doing
develops with the development of co-operation. For example, the continuous improvements of the 'learning curve' in complexes which use
just-in-time organizational principles together with an increase in
transactions and external linkages demonstrate forcefully the productive and innovative powers generated through this kind of division of
labour (see Swyngedouw 1992). These forms of co-operation through
externally articulated processes are superior to the sum total of the productive force embodied in each individual unit (Preteceille 1976). This
additional force is the combined result of what constitutes spatial
organization.

The components of the externalized configuration over which the
struggle for appropriation is fought can now be systematized:

(i)   the 'natural' goods; although already transformed through
      human interaction and whose productive powers are developed
      through the particular organization of the mode of production

and transformed through a labour process: natural resources, clean air, water, and the like;

(ii)   'collective goods': public fixed collective capital, comprising a) collective equipment necessary for the production and circulation process, b) collective equipment necessary for the reproduction of the labour force, and c) the ensemble of qualities of the labour pool;

(iii) private fixed collective capital, which includes the ensemble of existing capitalist units, resulting from the materialization of private capital to organize the production and circulation process (physical and organizational infrastructure);

(iv)   the 'institutional' and 'regulatory' forms; these are the (formal or informal, institutional or codified) practices, which generate, regulate, support and maintain the cohesion of the territorial configuration. They assure the reproduction of the configuration as well as its continuous transformation. These socialized practices equally regulate the mode of internalization/ externalization of the circulation process, but also express and (try to) contain the struggle over the appropriation of the effects of collective interaction. These forms can take a private character (firms, R&D labs, growth coalitions, chambers of commerce, technology or marketing organizations and so forth) or a public form (the state operating at various scales of territorial configuration, from the international realm to the neighbourhood level).

Each of these four sources of external capacities associated with territorial organization includes both quantitative elements (volume of resources, of fixed capital, of labour force) as well as qualitative elements (level of resources, of collective equipment, technologies, know-how, networks, institutions, inter-firm and other linkage patterns, qualification structures, and so on).

4.   Determining success

In the previous pages, I have argued the following:

1 Continuous innovation is the dominant competitive strategy in the current global competitive climate.

2 External relations and transactional structures are paramount in maintaining dynamics of permanent innovation in a context of inherent uncertainty and growing costs of internalized innovation.

3 Externalization presupposes forms of co-operation that may be problematic in terms of the necessary monopolization of innovation in a market environment.

4 The innovative capacity of transactional structures as well as the mediation of the collective/private dichotomy is structured through the institutional-regulatory and territorial 'embeddedness' of the transactional structure.

In brief, the institutional regulatory basis of the external relations is crucial both in generating an innovative dynamic and in mediating the tension between co-operation and exclusion (monopolization). Consequently, the capacity of different territorial configurations (or innovative milieux) to generate and diffuse innovations will depend on their respective transactional configurations. These configurations are essentially mediated through the following set of institutional regulatory practices:

- the organization and regulation of the capital-labour relationship
- the role and function of the local, regional and national states and their articulation
- the nature and codification of inter-firm relations
- the nature and codification of institutional relationships between firms and mediating agencies (research centres, development agencies, technology centres and the like)
- the regulation of resource allocation and in particular of capital
- the structure of innovative networks
- the ideological-cultural attitudes of political and economic actors.

These elements permit the exploration and analysis of the material basis of transactional networks and their innovative capacity on the one hand and the assessment of their success in mediating the collective/private antagonisms on the other. It is these characteristics which substantiate the functioning of externalized relations and which may account for the relative success of particular territorial configurations in generating and sustaining innovative capacity and, consequently, in assuring the sustained competitive dominance of others. What matters is the maximization of co-operative transactions and transcending or containing internal competition in a strategy to engage in inter-territorial competition. Or, put simply, 'local' co-operation is the basis for global domination. These elements will be explored further in the chapters of this book and particularly in the case studies. Sectoral and national or regional differences in organizing transactional configurations, particularly between firms and research/education centres, will be a key focus. This will permit the identification of a set of explanatory mechanisms resulting from the interaction of political and economic strategies, which underlie and shape the process of permanent innovation and innovation diffusion, on which successful geographical economic competition rests.

2
The Regulatory Context: International, National and Regional Components

1. Introduction

In the last two decades, throughout North America and Europe, industry and academic links have been identified as an important means for increasing output from national technical resources and improving industrial competitiveness (see David *et al.* 1994; Roobeek 1990; Ergas 1993; Charles and Howells 1992; Van Dierdonck *et al.* 1990, Chanaron 1989, and Brunat and Reverdy 1989). Official statements from each of the three case study countries and the EU express the view that the PSSB should stimulate and support innovation in industry. Examples from the UK include ACARD, 1983; ACOST 1991; Department of Trade and Industry (DTI) 1988; the 1993 White Paper 'Realising Our Potential'; from France the Ministère de la Recherche et de la Technologie (MRT) policy statement (1990), and from Belgium the Conseil National De La Politique Scientific (1984) policy document. Public policy in effect aims to create a set of norms whereby industrial access to research in universities and more recently, in national laboratories, is standard practice.

The purpose of this chapter is to examine some of the mechanisms by which norms of interaction are established at different geographical scales by public policy. It begins by discussing the reasons for political intervention in the process of innovation. Further sections discuss the specific forms of intervention adopted by the EU and by UK, France and Belgium governments, both at national and regional levels, designed to increase interaction between the PSSB and firms in the two sectors. The last section summarizes the main similarities and differences between policy and practice in the three countries.

34

2. Political intervention in the process of innovation

2.1. National systems of innovation

The concept of innovation filiere has as a central element that it is subject to pressures for change resulting from changes in government regulation. This is a variant of the innovations systems approach in which interdependence and interaction between the elements which comprise innovation systems is one of the most important characteristics (Edquist 1997, 21). These systems may take the form of *national systems of innovation* (see Lundvall 1992, Nelson (ed.) 1993; Edquist 1997) although studies which use this term recognize its limitations. Indeed, Nelson and Rosenberg (1993, 4) question whether in a world where technology and business are increasingly transnational, the concept of national systems as a whole makes sense. However, they pragmatically define a system as being 'a set of institutions whose interactions determine the innovative performance... of national firms'. These include public and private sector institutions such as industrial research laboratories. The behaviour of these institutions is influenced directly by explicit technology policies, indirectly by a range of other regulatory policies, and the mix of participating organizations, forms of ownership of information and plurality of regulatory form.

Technology policy has been defined by Stoneman (1987, 4/5) as 'a set of policies involving government intervention in the economy with the *intent* of affecting the process of technological innovation'. The argument for government intervention in the process of innovation through technology policies is made by Roobeek (1990, 14). She argues that innovation potential hinges on the capability to combine and integrate developments from diverse sciences and sections of industry and that as result, economic growth has come increasingly to rest on systematized technological and scientific research. Thus the state in principle has a role to play in

  (i)   overcoming industry's short-term horizons
  (ii) generating external effects arising from an extensive scientific infrastructure
 (iii) subsidizing very expensive but strategically important research
 (iv) gearing science and educational systems to meet current needs
  (v) overcoming problems of non-allocation for fundamental research arising out of failures of market forces.

Market failure also exists when there is the under-use by industry of public sector research. In a totally free market, all research would be conducted in the private sector. However, a private enterprise economy fails to provide adequate incentives for knowledge generation – which creates the need for public sector provision and the means of access. Insufficient access is undesirable because social returns from basic research are significant and higher than private returns production (Rosenberg 1990, 165). The market mechanism therefore has a tendency to discourage the production of public goods because of an inability on the part of producers to appropriate fully the value of the fruits of their efforts (David and Foray 1994, 29). These authors identified three generic remedies to overcome the deficiency of production in the market (David and Foray 1994, 13):

A Public production – where government engages itself directly in the production of knowledge e.g. through government research laboratories which disclose their findings.

B Private property and markets – granting intellectual property rights to private producers for their discoveries.

C Subsidies, procurement and regulated private production – encourage private production of knowledge by offering subsidies for its production, and by relying upon general taxation to finance these subsidies.

The argument is therefore, that the agencies responsible for encouraging industrial innovation have a role in facilitating the production of industrially useful information in the PSSB while stimulating demand. Moreover, actions which result from such intervention reduce the risk to industry of the uncertainties of longer-term research investment (see Rosenberg 1990, 165).

2.2   Scales of delivery of regulatory frameworks

Erik Swyngedouw argued in the previous chapter that collective intervention, including that by the state, can encourage the embedding of innovative activities in particular territories. The extent to which regulatory intervention is designed to facilitate localized and/or spatially dispersed information flows varies by country. Significant national variables include the defined parameters of access and the spatial balance of policy and resources allocated as innovation support strategies. These reflect differing priorities based on industrial heritage, politico-ideologies philosophies, and the location of economic power (Moulaert and Willekins 1987, 319). However, there is a general emphasis, even in the UK which does not have a regional tier of government, on the region as the source of national competitiveness.

The growing literature on global-localization (see Cooke *et al.* 1992; Gertler 1992; Ettlinger 1994) emphasizes the potential of regionally based delivery systems. Throughout Europe there are numerous examples of regional authorities fashioning new approaches designed to overcome the problem of low innovation potential (Cooke 1993) or to strengthen existing concentrations of expertise. Examples of the former include South Wales (Cooke 1998) and of the latter Grenoble (de Bernady 1998) and Catalonia (Escorsa *et al.* 1998). It has been argued that the absence of a regional tier of government places Britain at a serious competitive disadvantage (Cheshire *et al.* 1992).

While there are reservations about the ability of national governments to overcome barriers to technology transfer between industry and the PSSB, there is some scepticism about the possibility of effective regional or local intervention. Lipietz (1993,16) doubts that there is a margin of manoeuvre for regional social blocs which would be capable of putting into practice methods of local governance independent of national or continental politics (the EU for example) and of the world macro economy) and about a region's ability to influence economic development in general. Hilpert (1991, 21) argues that 'regional innovation policies cannot replace national policies for techno-industrial innovation'. A further aspect of the state as a facilitator is the extent to which institutions and technology transfer mechanisms are able to address the needs of firms at particular times for particular purposes.

To sum up this section, regulatory intervention to encourage interdependence and interaction between industry and the PSSB in theory does so by:

- influencing the kinds of information produced and available to actors at different spatial scales
- determining ownership, hence commercialization rights
- providing the means of formal access to information
- encouraging the development of norms of interaction which include informal relationships
- establishing frameworks for territorially based competitive strategies based on collective innovation relationships designed to strengthen or develop concentrations of expertise.

3.   European organizing frameworks

3.1   European Union policy

A powerful influence on national strategies in Europe is the growing EU budget for the support of R&D and in particular collaboration between the public research base and industry. (Since the coming into force of the Maastricht Treaty in November 1993, the European Community has become a component part of the EU, and legally there remains a difference between the two. In certain contexts the difference is an important one. However, for general purposes the 'European Union' is used to refer to what used to be known as the European Community (Bainbridge & Teasdale 1995). 'European Union' is used here to refer to what used to be known as the European Community.) Membership of the EU (and international agencies not focused on Europe) involves both the tendency to 'simultaneously internationalize and decentralize or devolve key policy issues' (Swyngedouw 1994, 5). However, the direct impact of EU initiatives is mediated by criteria for access and by industrial practices within each member state. These include national conditions of support for innovation, the resources available for firms to be even to be in a position to apply for funding, the attitude of larger firms to including smaller firms in proposals and exclusion of user firms in EC research programmes (Charles and Howells 1992).

At the beginning of the 1980s the Commission began to develop *Framework Programmes for Research* (Skoie 1995, 17) in order to strengthen the European science and technology base. But, as Skoie points out, it was not until 1987 that the European Community had a formal legitimate basis for handling matters related to R&D. In connection with the new *Single European Act* a new 'Title VI:Research and Technological Development' was introduced. Article 130F states:

> The Community's aim shall be to strengthen the scientific and technological base of European industry and to encourage it to be more competitive at an international level. In order to do this it shall encourage undertakings, including small and medium-sized undertakings, research centres and universities in their research and development activities; it shall support their efforts to co-operation with one another, aiming, in particular, at enabling undertakings to exploit the Community's internal market to the full, in particular through the opening up of national public contracts, the definition of common standards and the removal of legal and fiscal barriers to that co-operation (quoted in Skoie 1995, 17).

The Framework Programmes were mentioned specifically in the new Treaty. Skoie comments that these represent a centralized planning approach, although there has been no direct attempt at co-ordination of research policy in individual member states, the introduction of the single European Act in 1987 set the basis for such co-ordination. However, the official line is that the EU's research effort has been developed in conjunction with member states' research interests (Antonio Ruberti, EC Commissioner responsible for research and education, appointed in 1993, for the Fourth Research and Technological Development Framework Programme, (XIII Magazine 1/93, No. 10, 3).

3.2 European Union funding

Programmes are funded through the EU's regular budget and all member states contribute financially to the programmes. A total amount for R&D is allocated and decided upon within the EC's total budget and distributed according to the number of categories in each of the Programmes (Skoie 1995, 33). The scale and scope of the EU's support for research has continued to grow but remains a small proportion of total expenditure on R&D in the EU as a whole; in 1993 EC funding represented less than 4 per cent of total government research funding in the 12 member states. Tables 2.1 and 2.2 show the budget for the second and fourth Framework Programmes. Examples of European programmes which support innovation in the electronic components include ESPRIT and EUREKA. The BCR (now the Standards, Measurement Testing Programme) funds research into metrology.

Table 2.1   Second EC Framework Programme on research and development in enabling technologies

Action lines	MECU
Enabling technologies	
1. Information and communications technologies	
Information technologies	1352
Combination technologies	489
Development of telematics	380
2. Industrial and materials technologies	
Industrial and materials technologies	748
Measurement and Testing	140

Source: Charles and Howells 1992, 105

Table 2.2   Fourth EC Framework Programme on research and development in enabling technologies

Field	Funding MECU
1. Information and communications technologies	3405
Information technologies	1932
Communication technologies	630
Telematics	843
2. Industrial and materials technologies	1995
Industrial and materials technologies	1707
Standards, measurements and testing	288

Source: *Measurement and Testing* 1994

Two important issues arise from the allocation of funds between components of the programme and the distribution of awards between member states. The first is whether EU funding is complementary to or a substitute for national funding. The second is how the construction of national innovation systems affects the type of participant in EU programmes.

The case for EU funding to be complementary to rather than substitute funding for national funding has been developed by David, Guena and Steinmueller (1995, 11). They propose an economic theory of the aims of additionality as a principle in the allocation of EU funds in a study for the STOA programme of the European Parliament. They argue that 'a goal for EC RTD funding is that it should create benefits that are greater than (additional to) those that would be available through a strictly national allocation of RTD funds within Europe'. Their report specifically addressed two questions. First, what EU principles will insure that EU funding of RTD will not be a replacement or substitute, for the efforts of member states? Second, what will be the response of member states, and the institutions within those member states, to the potential or actual receipt of research funds from the EU? (David *et al.* 13). They conclude that patterns of research specialization may be similar for both EU and member state RTD funding. However, their study found that generally substitution is neither a systemic or pervasive characteristic of member state response to the receipt of EU funds. In one country and one programme a substitution relationship was found. This was the UK in the case of the Brite-EURAM programme. They found, 'that for larger institutions of higher education,

Table 2.3 Distribution by participant types in those parts of the Second Framework Programme for the UK and all countries

Organisation Type	United Kingdom (per cent)	All countries (per cent)	*All countries (including DGXIII) (per cent)
Large firms	13	11	15
SMEs	9	12	15
Research organizations	29	37	31
Higher Education institutions	49	37	35
Other	1	3	3

Source: Georghiou *et al.* 1992
*Column 4 gives the same distribution for the whole of the Second Framework Programme, including DGXIII programmes but excludes BCR.

EU and national funding are likely to be complementary while for smaller universities and university departments there are indications that substitution is important' (p. 47).

The construction of national innovation systems has been shown to have an effect on the type of participant in EU programmes. Table 2.3 shows proportion of participants by each type of participant in those parts of the second Framework Programme managed by DGXII and by DGXIII. It indicates that the UK, compared to other countries, is slightly more likely to be represented by large firms than by SMEs, possibly because there are more large firms in the UK than the average for the EU. The very high representation of UK HEIs partly reflects structural differences in the location of research and partly the commitment to participate by UK HEIs, and the greater proportion of research in HEIs than in research organizations (Georghiou *et al.* 1992, 16). Moreover, collaboration has become so popular with UK firms that the UK has the largest number of collaborative links of any member state (Georghiou *et al.* 1992, i).

3.3 Specific programmes in electronics and metrology

ESPRIT

ESPRIT was launched in 1984, virtually at the same time as the UK's ALVEY programme of advanced IT research. ESPRIT I was the result of action by the EC in response to a call by the so-called 'Round-Table' of

12 leading European IT companies. The aims of the programme were to underpin the technology base of the IT industry, foster collaboration, and assist in the standards area. ESPRIT I (1984–8) was followed by ESPRIT II (1988–92) and ESPRIT III (1992–8). Projects, as in all EU ICT programmes, must as a rule involve organizations from two member states; and the basic level of Community support is 50 per cent of project costs. ESPRIT III covers six technical areas:

   (i)   microelectronics;
  (ii)   information processing systems and software;
 (iii)   advanced home and business systems and peripherals;
 (iv)   computer integrated manufacturing and engineering;
  (v)   basic research actions;
 (vi)   open microprocessor initiative.

By May 1993, ESPRIT had involved over 1500 participants in 600 projects (European Supplement to JFIT News Issue No.43 May 1993). However, after ESPRIT II, the 'Dekker Report' argued that the programme should not continue in its present form. ESPRIT's industry-led, technology-push approach was capable of producing good results but was not achieving the lift in European competitiveness that was called for. Instead, the Report recommended that the broad focus of ESPRIT should be narrowed. To clarify the focus, a set of smaller, interrelated programmes, each with just a few key objectives, was advocated. For example, user needs should receive greater emphasis.

EUREKA

The EUREKA initiative, not part of EU collaborative research programmes, was created in 1985. Its aim is to improve cross-border collaboration in research and development projects to strengthen the productivity and competitiveness of European industry in world markets (EUREKA-INTO NEWS December 1994). EUREKA constitutes an umbrella mechanism for encouraging firm-to-firm collaboration. It applies to all Western European countries, not just the EU, and has no central funding. EUREKA recognized projects qualify for national R&D support funding, the terms of which vary from country to country (Sharp 1991, 71). Partners negotiate their own intellectual property right (IPR) arrangements, unlike in EU programmes which have formal IPR guidelines. In 1994 EUREKA had 22 members including Austria, Belgium, Denmark, Finland, France, Germany, Hungary, Iceland, Ireland, Italy, Luxembourg, The Netherlands, Norway, Portugal, Spain,

Sweden, Switzerland, Turkey, the UK and the Commission of the European Communities.

The EU and EUREKA joined forces for Joint European Submicron Silicon Initiative (JESSI) to fund the programme jointly with large electronics firms and national governments. The JESSI programme was scheduled to run from 1988 to 1995, with funding ending in 1996. The aim of the programme was to match US and Japanese efforts in developing technologies for submicron features that were expected to mark high-density memory and logic chips in the 1990s. JESSI included basic research, development of product technology, circuit applications and a focus on materials and equipment including a pilot production line. Over half of the $2–$4 billion total investment was scheduled to be financed by participating European governments. The emphasis is on application-specific integrated circuits (ASICs) because of their wide applicability. The programme has its own rules for inclusion and exclusion. For example the British computer firm ICL was initially excluded from the JESSI initiative after it had been taken over by Fujitsui.

JESSI has been judged to be of major importance to the European electronics industry, delivering a half micro logic process to its commercial partners, which included both major and smaller electronics companies. It has 'brought European microelectronics back from the brink of being overwhelmed by Japanese and American Competition' (Manners 1994).

BCR and Standards, Measurement and Testing Programmes

BCR was part of the Industrial and Materials Technologies Programme. It was allocated only 140 MECU under the second Framework Programme. BCR was replaced by the Standards Measurement and Testing Programme on 15 December 1994. Under the Fourth Framework Programme the Standards Measurement and Testing programme's budget at 288 MECU was double that under the Second Framework Programme, but as the budget for the whole Industrial and Materials Technologies Programme had more than doubled, BCR comprised an even smaller component of the programme than a decade earlier.

4. National investment strategies

4.1 National differences in funding of R&D

The proportion of GDP of government to industry funded university research, as well the allocation of funds to support interaction, is an important determinant of the extent to which industry and the PSSB interact. This is influenced by national characteristics of rewards to shareholders which in turn influences the relationship between corporate structure and R&D strategy (Malecki 1990, 162). Table 2.4 shows that there are considerable differences between the three countries in R&D expenditure by source of funds as a percentage of GDP for Great Britain, France and Belgium for 1990.

The major differences between the three countries are in levels of overall spending, the proportion spent on defence and the degree of industrial versus public spending. France allocates the highest share of GDP to R&D and Belgium the lowest. The main difference between the UK, France and Belgium is the proportion of defence as a percentage of total R&D. In the UK and France, both industry and government have traditionally allocated a substantial share of R&D expenditure on defence related activities. The UK has the second highest proportion of government outlay for military R&D expenditures (45.2 per cent) of OECD countries behind the US (58.6 per cent). France has the third highest (34.6). Belgium spends very little on defence R&D. Industry finances the largest proportion of R&D. However, expressing expenditure in percentage of GDP suppresses differences in the actual amounts available for R&D and innovation. To show these differences more clearly, Table 2.5 gives the expenditure per capita of the workforce in constant value £(1991) using OECD purchasing-power-parity (ppp) currency exchange rates.

This shows that compared to the UK, industry in Germany and the USA was spending more than twice as much on developing new ideas and technology for every member of the labour force. France was spending 20 per cent more than the UK, but was still short of the US, Germany and Japan. In the year 1990 to 1991, UK industrial spending

Table 2.4   R&D expenditure by source of funds as a percentage of GDP in 1991

	per cent GDP	Publicly funded R&D as per cent of total R&D	Defence R&D as per cent of total R&D
Great Britain	2.1	35.5	44.8
France	2.4	48.8	36.1
Belgium (1990)	1.7	27.6	0.2

Source: OECD in Skoie 1995

Table 2.5 Industrial funding £(1991) per capita of the workforce using OECD ppp currency exchange rates

Year	UK	France	Belgium	US	Japan	West Germany
1981	153	172	n/a	290	233	297
1990	205	261	n/a	382	470	412

Source: Save British Science 1994, 4

on R&D fell by 10 per cent in real terms, while in France during the 1980s the percentage increased (Save British Science 1994, 4).

5. The UK system and policy

5.1 Policy

The UK's research effort was characterized by 'big science deployed to meet big problems' and by the concentration of R&D effort in a small number of technologies of strategic importance (Ergas 1989, 52). During the 1970s for example, in the industrial sector, technological resources became excessively focused on high technology projects (for example Concorde and the Advanced Gas-cooled Reactor) to the detriment of sectoral performance (Walker 1993). In the past therefore, technology policy in the UK, as in France and the US, was, Ergas argues, 'intimately linked to objectives of national sovereignty'. This strategy invested in the major funding agencies a considerable amount of discretionary power, which Ergas suggested inevitably benefited the country's major firms.

The focus of criticism in recent years has turned to the state of the UK science base. Concern has been expressed from within the political system and from academia. Criticisms fall into three main categories.

A. That science has been under-funded

This is reflected in criticism expressed in parliamentary reports such as those by the House of Lords Select Committee on Science and Technology and by the House of Commons Select Committee and by those produced by the academic pressure group 'Save British Science' (SBS).

The problem of under-funding arises from the UK's recent policy of exploiting rather than only developing technology. This philosophy was laid down in the 1988 White Paper 'DTI – the department for enterprise'. The White Paper contained three central concepts (Walker 1993, 184). One was 'open market', which in practice meant that 'competition policy' came to form the core of the government's industrial policy. The second concept was 'enterprise' which gave the market its energy and creativity. Lack of enterprise was identified as having played a major part in the relative decline of the British economy (DTI 1988, 1). The third was that the government's support for innovation should be constrained in expenditure terms. The emphasis should be placed in achieving greater value-for-money by raising the efficiency with which resources were used, not least by making their allocation conditional on recipients satisfying strict performance criteria. Walker argues that this meant that in relation to science, to education, to R&D, and all other areas where the state played a part in the innovation system, assessment of performance was based on demonstrable returns, and that long-term projects with uncertain paybacks were rationed. Walker therefore concludes that the government's attitude towards science and technology prioritized productivity and cost reductions before expansion and creation of new capabilities, even where, as in education, there was a historic tendency towards under-investment.

According to Walker (1993, 186) what has been lost in the move from the 'mission-oriented' approach that Britain shared with France and the United States, to the 'diffusion-oriented' approach, has been the recognition of the 'importance of building strong technological capabilities'. This has manifested itself by the country's failure to invest in strategic areas. The UK has neither 'shown little appetite for the investments required for the heavy investments required to maintain strong indigenous industries' nor been willing 'to play a part in identifying and supporting technologies which have a strategic value'. Examples of lack of commitment to supporting a domestic electronics industry are the sale of INMOS to SGS-Thomson and ICL to Fujitsui. They illustrate Britain's 'unconcern over the fate of its remaining semiconductor and large scale computer capabilities'.

The move away from the 'mission-orientated' approach has affected the structure of the electronics industry in three respects.

1 The UK government prefers to support inward investment in electronics (Manners 1994) rather than build a strong technology base. Inward investment in production has gone hand in hand with an increasing share of R&D conducted by foreign-owned firms.

Overseas funding of industrial R&D more than trebled over a little more than 20 years from 4 per cent in 1967 to 13 per cent in 1986 (Charles and Howells 1992, 15).

2 Part of the problems facing the UK computer industry is related to the failure of the government to encourage firms to exploit research developed under the ALVEY programme, due to a misunderstanding of what R&D alone can achieve. Particularly significant was the absence of training initiatives to address skill shortages. Their technological weaknesses have made them more vulnerable to take-over (Oakley and Owen 1990).

3 The UK has been reluctant to pay its share in supporting JESSI. Britain's contribution is about the same as Portugal's and is about a tenth of what the Netherlands contributes, which is less than half the individual contributions of Germany and France (Manners 1994).

The election of a Labour government on 1 May 1997 brought a greater commitment to funding science and technology. The Comprehensive Spending Review announced on 13 July 1998 contained an allocation of £1.4 billion new money, including £400 million from the Wellcome Trust. Most of the money (£700 million) was to go to the Office of Science and Technology, essentially the Research Councils (SBS Winter 1998, 1).

B. That the state has not acted as the catalyst of industrial and technological development as it has in France, Japan and several other nations

The failure of the state to catalyse developments is caused by two features of the innovation system. First, science and technology policy in the UK has traditionally been decentralized, with individual ministries responsible for their own research programmes which they commission to support their particular missions (Georghiou *et al.* 1992, 11). Overall advice on science and technology issues is provided by the Advisory Council on Science and Technology (ACOST). In recent years there has been a tendency towards centralization. In 1992 responsibility for co-ordination was brought together with the creation of the Office of Science and Technology (OST) in the Cabinet Office, headed by the Chief Scientific Adviser. The Office has responsibility for the Science budget. This includes support for the five research councils (see below). The Chancellor of the Duchy of Lancaster then provided science and technology, as a single entity, with its first cabinet representation since the 1960s. Overall cabinet responsibility remained with the Prime Minister. A further aspect of these changes was the inclusion in the new Office of the responsibility for the Science budget. This responsibility was short-lived. In 1995, the Department of Trade and Industry (DTI) took overall control of the Science budget. While the move of the control of spending on science to the OST within the Cabinet Office 'brought it to a more central position in the government and increased its importance to decisions in almost all sections of policy' the removal of the Office to the DTI was seen by SBS as being 'reduced to the ranks' (SBS Autumn 1995, 1). The change meant that instead of the OST being engaged in a constructive dialogue with a Ministry that SBS believed was listening, it was 'banished to a corner of the DTI, a department with an appalling record in the understanding of the research process and its management'.

The second aspect of the innovation system which affects this is, more generally, the operation of the financial system in the UK which in turn reflects a particular kind of regulatory regime. UK firms' ability to benefit from whatever level of public spending is influenced by spend on in-house investment in R&D compared to returns to shareholders. UK companies tend to pay out more in dividends than companies in principal competitor countries. This may be a reflection of an attitude of 'jam today rather than investment for the future' coupled with a lack of appreciation that improving practice reduces the risk of failure, increases confidence for further investment and generates new business (DTI/CBI 1992, 7). The operation of this system is very different to that of Germany where the incidence of industry and academic links has been furthered by access of several smaller enterprises to funding from the Lander (provincial) governments (Mason and Wagner 1994). There banks provide 'patient capital' and guide the development of important sectors.

C. The weaknesses in the national innovation system are the lack of integration of scientific and technological communities and the lack of co-ordination between the different elements (Walker 1993, 180/182)

The lack of integration was addressed in the 1988 White Paper. This spelt out that industrial competitiveness was to be achieved by encouraging collaboration between HEIs and companies (at the pre-competitive stage). The UK's overall weakness in industrial innovation performance has long been identified as arising out of its scientific capability into commercial application. Recent philosophy was presented in a report by the Parliamentary Office of Science and Technology (Webster 1994, 191). This suggested two key doctrines' driving reform of the public sector research base. The first (markets) follows from ideas of 'new institutional economics' which seek to give service users greater choice by constructing internal and external markets, increasing competition, privatizing and contracting out. The second (managerialism) is the pursuit of business-type professionalism in the public sector by decentralizing services, and giving managers delegated powers to achieve the most efficient output. The introduction of these doctrines into the academic sector is indicated by market systems developing through commodification of academic research through patenting and business services.

5.2 Allocation of resources

(The information on allocation of resources for both the UK and France is taken from Atkinson *et al.* 1990 Volume 2.) R&D performers in the UK fall into five categories. The first four comprise the UK's domestic performers.

(i)   government
(ii) higher education
(iii) industry
(iv) others
(v)   abroad (UK government-funded research undertaken abroad, for example CERN for particle physics).

Government

The government category is divided into three groups: those of the Ministry of Defence, those of civil Ministries, and those of Research Councils. Each of these to a greater or lesser extent, except the Economic and Social Research Council, also funds research in other organizations. The Advisory Board for the Research Councils (ABRC) advises on the distribution of funds to the research councils.

Funding

In the UK, there is no research ministry or overall research budget as there is in France. Academics can bid to the five research councils for funding, to charities, and other government departments such as the MoD and the DTI. In 1987 together the research councils spent intramurally about £350m. The five funding councils were:

• the Agricultural and Food Research Council (AFRC)
• the Medical Research Council (MRC)
• the Natural Environment Research Council (NERC)

- the Science and Engineering Research Council (SERC)
- Economic and Social Research Council (ESRC) which carries out no intra-mural research.

In 1994 SERC was divided into the Engineering and Physical Sciences Research Council (EPSRC) and High Energy Physics. The AFRC, MRC and NERC all fund their own research institutes. Under the old system SERC had four research establishments, whose function was the support of researchers in the HEIs through the provision of major research facilities. SERC also funded research in specialist centres in universities, such as the micro-fabrication facility at Edinburgh University.

While the universities are the biggest performer of basic and strategic research in the UK, industrial laboratories collectively are the biggest research funders in the UK. Industry includes private industry, public corporations (including what were then the laboratories of the United Kingdom Atomic Energy Authority (UKAEA)) research associations and industrial research institutes. Research (then about £350m a year) is also performed by other organizations such as non-industrial research institutes, e.g. those of certain charities, and professional and learned societies, for example, the Royal Society. Some research is performed in international research organizations mostly located abroad, for example CERN for particle physics. Overall, the biggest performer and funder of R&D is the MoD. In the late 1980s this amounted to over £2 billion a year, a significantly greater amount than the £350m spent by the research councils.

Innovation support for industry and for technology transfer

The DTI and other government departments encourage and finance collaborative research in four main ways:

A. The LINK programme   This is designed to encourage companies to undertake joint research with HEIs and research councils. It provides government support of up to 50 per cent of eligible costs for strategic research, involving companies and science institutions in collaborative projects. It involves all major departments of government (for example Health, DTI) and the research councils. From September 1993, LINK became the only national collaborative R&D mechanism (Eddison 1993,11).

B. Support for the EUREKA programme   National collaborative research programmes promote longer-term industrially led collaborative projects between UK companies in advanced technologies. For example the DTI, with advice from its Technology Requirements Board, was running a collaborative programme in gallium arsenide, an important semiconductor.

C. General industrial collaborative projects Examples include Collaborative Awards in Science and Engineering (CASE) funded by SERC, and the Teaching Company Scheme, jointly supported by SERC and the DTI.

D. Support for high technology small firms This takes place through the Small Firms Merit Award for Research and Technology (SMART). The new SMART scheme introduced in 1997 combined the previous SMART and SPUR (Support for Products under Research).

Industry-specific programmes

In electronics, the ALVEY Programme constituted the single largest UK IT policy initiative in the 1980s costing the public purse some £200 million, with industry contributing an estimated £150 million. One of its main aims was to improve the competitiveness of the UK IT sector. The programme began in 1983 and was sponsored by the DTI, SERC and MoD. One of its main achievements was in fortifying the research community, and in nurturing links between academia and industry (HMSO 1991, i). Although LINK was intended to bridge the gap between industry and academia, at its inception it was anticipated that it would be unlikely to generate the gains of ALVEY in fostering closer links between the industrial and academic communities because it was another fragmented initiative (Georghiou *et al.* 1992).

More recently, a number of industry/academic programmes and industry-only Advanced Technology Programmes (ATP) initiatives aimed at key themes in opto-electronics and microelectronics have been introduced. However, economic restraints severely limited the number of projects which could be funded. The consequent need to prioritize areas for support has led to the early closure of a number of programmes.

Academic research in electronics is co-ordinated through the seven Information Technology Advisory Boards of the Joint Framework on Information Technology (JFIT). JFIT was established in 1987 by the DTI and SERC to 'co-ordinate their support for advanced IT research, technology transfer and education and training, with the aim of providing greater coherence in the Government's support for civil IT research'.

Under JFIT are a number of initiatives: LINK, Advanced Technology Programmes, the ALVEY Programme, JOERS and National Electronics Research Initiatives. Heavy investment by SERC (£80m starting in 1984) into low dimensional structures, occurred at a time when UK industry began to cutback on this area. The SERC had since abandoned its silicon committee. EPSRC provides funds for academic only programmes such as Silicon Devices and Processes (SIPD) and Novel Systems on Silicon (NSOS). The NEDC had recognized the gap between industry and universities, and recommended that the Government could do more to promote collaboration by pump priming joint academic/industry projects (1991,6). To this end it initiated the Collaborative Research and Development in Electronics (CRADE) project designed to bring universities and industry together. This ended with NEDC's demise in the early 1990s.

One LINK programme was directed to the flow measuring industry. This was the (Joint) Industrial Measurements Systems Programme (JIMS). It was jointly funded by the DTI and SERC, was launched in April 1988 and ended in 1993. Initially £11m was allocated over six years for collaborative research in this field. By August 1990, 25 full proposals had been approved (35 full applications; 98 proposals commented on) (LINK newsletter, August 1990). The programme concentrated on the development of instrumentation and measurement technologies where there is emphasis on system integration. Priority topics were micro-electronics, optical and spectroscopic instrumentation, and sonics and ultrasonics. This was replaced by a new LINK programme TAPM.

Until 1 September 1993, Research Council research in UK universities for instrumentation was funded through the SERC's Control and Instrumentation Committee, within the IT directorate Joint Framework on Information Technology (JFIT). It was then moved to the Engineering Research Commission, and renamed the Integrated Control Systems Engineering Committee (ICSE). The committee funds research into techniques used within instrumentation such as ultrasound. Research into the use of instrumentation falls within the orbit of the process committee. Two broad categories of research are supported by the ICSE: control and instrumentation, in equal proportions. Funding in 1993/4 was higher than previous years: £6m compared to the usual level £3–3.5m.

Clubs

The organizational framework in which innovation takes place in the UK includes 'clubs' supported by the DTI. By October 1987, the DTI funded 104 clubs of different kinds, with some 2430 members. Of these, large firms (over 500 employees) constituted 63 per cent of UK members and only 16 per cent were SMEs, and 19 per cent of all members were foreign. They had 'a mainly pre-competitive role aimed at the diffusion and uptake of scientific and technical knowledge and best practice mainly with fairly long-term implications' (DTI 1988, i). An evaluation of clubs (DTI 1988) found that around three-quarters of members rated clubs very highly for research effort and dissemination. However, smaller firms had difficulty in devoting sufficient time to their membership, appeared to get less out of it and rated it lower. The UKAEA as a whole was host to 15 clubs (DTI 1988). The first club at Harwell was Heat Transfer and Fluid Flow Service formed in 1967. NEL runs the Flow Liaison Club which meets regularly, moving between different industrial and academic locations. In 1995, a new club, the NEL Multiphase Flow Club was formed. By April 1995, ten members, from the UK, US and Norway, had signed up (Flow Tidings No. 9 April 1995).

Regional organizations

Limited authority has been dispensed to the regions in the UK. The organizations with the most power and resources are Scottish Enterprise and the Welsh Development Agency. Other less significant moves have been the creation of local enterprise agencies (LEAs) supported by combinations of local government, business and financial institutions, in order to provide support facilities for small firms. By 1989, 300 had been established and an increased role for the regional DTI offices and the setting up of the 14 Regional Technology Centres. The latter have since either privatized or transformed into different organizational forms. In the early 1990s academic commentators felt that the spatial consequences of technology policy had been largely ignored by policy makers, the result being that regional disparities are likely to widen (see Charles and Howells 1992, Begg 1993). The Labour Government elected in 1997 has devolved power to the regions through the establishment of Regional Development Agencies on 1 April 1999.

6. The French innovation system

France is the fourth largest OECD economy, behind the US, Japan and Germany. Since World War II, the state has played a considerable role in shaping the industrial fabric of France, directly through the national planning agency DATAR and its ownership of most of the banking sector and some major companies, and indirectly through the close links and constant interchange between civil servants and much of France's top industrial management.

6.1 Policy

Atkinson *et al.* (1990, 4) describe the French research system as being a mirror image of that in the UK. There is a single ministry, the MRT, with overall responsibility, a minister who is in the cabinet and a substantial budget. Like the UK, France has a commitment to specific areas and to large scale science and technology generally. Basic and strategic research is the province largely of the MRT's national centre for scientific research, CNRS. Science has been a high priority in French policy for many years, 'It is espoused from the President of the Republic down'.

The French national system of innovation as a whole consists to a large extent of vertically structured and strongly compartmentalized *sectoral subsystems* often working for public markets and invariably involving an alliance between the State and public and/or private business enterprises belonging to the oligopolistic core of French industry. Electronics is one of the most important subsystems (the others include electrical power production, telecoms, space and arms production) (Chesnais 1993, 192).

The French system, like that of the UK, has been subject to a number of criticisms.

A. The initial failure of measures to promote exploitation of publicly funded research followed by an excessively well-developed relationship with industry

Some measures to commercialize public sector research have been in place since the early 1980s. For example, in 1981, the CNRS created a major division for the exploitation of science with the objective of promoting links with industry (Borde 1992). However, these links were slow to develop. By the mid 1980s, two thirds of CNRS laboratories had no ties with industry and only 1000 out of 80,000 firms had signed agreements with CNRS (OECD 1986, 208). This pattern showed a dramatic change. By the 1990s there had been a pronounced shift towards research contracted out to the CNRS to the extent that there was some concern that things had gone too far. The level and type of work being conducted in CNRS laboratories on behalf of the private sector had become so extensive that a recent review warned that, 'CNRS must be vigilant not to become a subcontractor to industry' (quoted in Webster 1994, 194).

B. The failure of defence R&D to be translated into commercial products in the civil sector

Chesnais finds that the evidence suggests that a large part of French high technology industry has been shaped by the pervasive influence of defence markets and military demand, notably the highly customized, non-generic uses of military technologies and their very low transferability to civilian uses. Moreover, in instances where new technologies emerge in the defence sector, as in laser technology, the transfer to civilian user has proved a complete failure.

C. Industrial performance and therefore the success of the French innovation system, measured by imports and exports, reveal a serious situation, with a long decline in exports, matched by a rise in imports (Chesnais 1993, 222)

This failure is not to do with the insufficient level of French R&D expenditures, as this is comparable with France's competitors. It does 'have a lot to do with the structure of industrial R&D, the *priorities* chosen, and the institutional context in which R&D is undertaken' (Chesnais 1993, 225). Part of this is the particular rigidities of the state-led system, in which large firms such as Thomson occupy central positions in policy making processes (see Hancher and Moran 1989, 289). Moreover, it has to do with the dominance of large firms as recipients of government research funds. Small firms as beneficiaries receive less than three per cent of government R&D subsidies (Chesnais 1992, 209).

D. A lack of enterprise in France

An OECD report published in 1986 found that in France fewer firms were created, fewer activities completed and a definite lag compared with the main foreign competitors (OECD 1986, 225).

E. The weakness of R&D

Industrial R&D, or R&D carried out within firms, remains significantly weak (Chesnais 1993, 208).

6.2 Allocation of resources

Atkinson *et al.* (1990, 28) identifies the four broad categories of research performers in France:

(i) industry
(ii) 'administrations' (comprising all public sector performers and universities);

(iii)   private non-profit (PNP)
(iv)   abroad.

'Administrations'

The public sector is divided into four types of performer, the first three on the civil side and the fourth for military.

(i)   public sector research organizations (OPR)
(ii)   public establishments of administrative character (EPA) and administrative services
(iii)   higher education
(iv)   military research.

The first is divided into:

Public establishments of scientific and technological character, EPST.   This category includes institutions funded by the MRT and which generally undertake basic and strategic research. This includes the National Centre for Scientific Research (CNRS).

Public establishments of industrial and commercial character: EPIC.   These organizations have a substantial commercial or industrial element integrated into the work. Part of the Atomic Energy Commission (CEA) falls into this category.

Funding

Chesnais (1993, 192), identified several features which are specific to the French system of funding public sector research:

(i)   the organization and funding of the largest part of fundamental research through a special institution, the CNRS, distinct from higher education entities, which are funded by the state and governed by scientists in an easy relationship with public authorities;
(ii)   a dual higher education sector producing at least one type of senior technical person little known elsewhere. The 'Grandes Ecoles' produce technical experts – the elite of engineers cum industrial managers cum high level political and administrative personnel;
(iii)   a pervasive element of state involvement in the production not just of general and scientific technical knowledge, but often of technology *per se* in the form of patentable and/or immediately usable products or production processes.

The government funds approximately 50 per cent of R&D and industry about 44 per cent, the rest coming from foreign sources. About 55 per cent is executed in industry and 43 per cent in the public sector which includes both government research laboratories (about 27 per cent) and higher education research (about 16 per cent), which includes funds provided by the CNRS and general university funds as reported in OECD statistics, which are of course basically aimed at teaching and only subordinately at research (Chesnais 1993, 206).

Policy instruments

France has a centralized system of 'consolidating and managing the entire range of resources'. At the time the study was undertaken, the MRT was responsible for the support of government-funded civil R&D, and for co-ordinating most of the rest through the mechanism of the civil R&D budget, the BCRD.

In the mid-1980s, the OECD (1986, 173) identified four procedures of direct aid to industry:

A. *The fonds de la recherche et de la technologie (FRT)* The purpose was to speed up implementation of France's research and technological development policy by providing financial incentives.

B. *Agency nationale de valorisation de la recherche (ANVAR) aid to innovation* See below page 60.

C. *Special investment loans (PSIs)* The purpose of these is to:
- renew the industrial fabric
- stimulate investment
- automation of SMEs
- stimulation of exports.

D. *Fonds Industriel de Modernisation (FIM)* This was an industrial fund set up alongside ANVAR to help manufacturing firms to finance investment in order to modernize manufacturing processes or develop new products and processes.

In addition a series of tax measures and measures to increase the availability of venture capital were in operation. In particular, there was a tax credit for research. This involved a reduction in firms' corporation or income tax liability equivalent to 50 per cent of the year-on-year increase in their R&D effort.

Sectoral support

France, like the UK, has a history of institutionalized support for the electronics industries. Innovation support generally is primarily through the research ministry (MESR) centrally and through its regional offices, and regional offices of ANVAR. The MESR's predecessor, the MRT, also supported the Industrial Conventions through Technological Research (CIFRE) projects – equivalent to the CASE scheme in the UK – and funds doctoral candidates. France also operates a research tax incentive which has the objective of promoting research and development by SMEs. It works by the government funding the increase in R&D spend from one year to another.

In the case of electronics, the argument for support was made in the preparatory report of the Electronics Sector Mission 1982. It stated that, 'If France is to maintain its independence, master the new communications systems and put the crisis behind it, it is imperative for the country to master the key activities of the electronics sector by the end of the decade... If France has to implement an overall strategy for a single industry, it should chose the electronics sector' (OECD 1986, 218).

The OECD Report identified the electronics sector as central for three reasons:

1 It forms a vast industrial entity including strategic activities such as components, the computer field, electronic office equipment, automotive systems, telecommunications, industrial and military electronics, consumer electronics and computer services. This entity has been identified as the world's main 'industrial growth pole' with a world production of $365bn in 1983 (3.5 to 4 per cent of world GDP and real world growth of 5 to 7 per cent).
2 France has an important place in this sector even if the situation varies with each sub-sector – 4.8 per cent of the market and 4.5 per cent of world production and a workforce of 300,000.
3 The government has been investing substantial funds in it over a long period.

The Report found that the competitiveness of the French electronics sector varies greatly. It is strong in telecommunications, industrial and military electronics and software, but electronic components and computer fields are marked by the great dependence of national firms and strongly placed multinationals. Local supply is limited in the case of consumer electronics, electronic office equipment and automatic systems. France was regarded as particularly weak in many up-market segments (micro-computer business, robotics, Computer-aided Design (CAD), software packages, video-recorders – those which will be expanding the most in the future. The action plan for the electronics sector (PAFE) was set up to achieve technological independence in integrated circuits and the computer field.

Two main actions were taken initially and had the biggest impact on subsequent developments.

1 In each sub-sector, the government set out to create one or two national champions with which the relevant activities of the other French electronics firms were incorporated. In the components field, the activities of Thomson, Matra, Radiotechnique, Saint-Gobin-Eurotechnique and LETI were merged with two firms, Thomson and Matra, each of which specializes in a technology (respectively CMOS and NMOS) (OECD 1986, 220).
2 An enormous financial investment. Thus FF 140bn were to be invested in R&D and productive investment under PAFE in the period 1983–7 by the French electronics sector from firms, the armed forces, government PTT and so on. FF 60bn was to come from government.

The 1986 OECD Report (pp. 221–2) makes five criticisms of PAFE arising out of its cumbersome approach.

1 Problems were created by merging firms instead of mastering co-operation/competition techniques which had been achieved in Japan.
2 National champions absorbing the bulk of government funds is not necessarily the most efficient use of funds and this method of support has detrimental consequences for small firms. Therefore general support for the industry is more effective.
3 The choice of a broad industry approach is often accompanied by a policy of national preference. In the case of France the government's protectionist attitude has slowed down or even obstructed the computerization of public research bodies and made it more expensive.
4 Some advances have been made in components and medium and large scale computers. It would be better if France was more specialized in which sectors it supported.

5   The consolidation of European initiatives in the last few years calls for greater specialization.

Products using Electronics Components (PUCE) was launched in 1983 and began operating in 1984. It was aimed directly at the user and future user in industry and provides financial support for industrial projects in which the introduction of electronics components or technology would be essential for the undertaking. It is aimed at any company with under 2000 employees. Although there were difficulties, firms using PUCE had improved their capability in micro-electronic technology (OECD 1986, 86).

Regional innovation support

France has a far more developed regional structure than the UK. The country is divided into 22 regions (created in 1972), each of which encompass several departments (92 in total). The directly elected Regional Council, and an Economic and Social Council in the regions 'serve chiefly as focal points for channelling government funds to different departments and, increasingly, to encourage inward investment in their area' (DTI 1991, 4). The Lisle (1973) and Delion (1974) Reports demonstrated that from the Second Plan to the Sixth Plan (1971–5) the state was conscious of the role of science and technology in economic growth, and that as such, funds should be made increasingly available for R&D across the set of regions. The Seventh Plan (1975–80) sought to put these beliefs into practice (Buswell 1983, 9). ANVAR was one of the most important organizations set up under the Seventh Plan.

ANVAR is a regionally based organization. It was established in order to co-ordinate the vast operation of transferring technologies downstream (OECD 1986, 209). The organization has two functions, a) to develop research and promote innovation by providing direct aid to innovators using banking services and their specialized services to assess the financial risk element; and b) the task of modernizing industry. For this ANVAR calls upon the resources of FIM. In the mid-1980s ANVAR had 24 regional offices and a staff of 500. The CNRS works closely with ANVAR through its regional organizations. It has contact with almost 5000 firms, manages a portfolio of patents for the major bodies (1004 patent applications in France and abroad), and has a good sectoral coverage mainly involving SMEs, particularly through the procedure aid to innovation. It has other functions connected with the use of services, familiarizing young people with innovation, undertaking feasibility studies and providing regional supplements to innovation aid. According to the Ministry of Industry documentation, one franc invested in ANVAR is converted to six francs in industry (OECD 1986, 174).

DRRT serves as the national configuration of research at the regional level. The DRRT represents the Ministry of Research and Technology in each region. Its function is to develop co-operation with organizations within regions and attracting others which would strengthen the fabric of the region. Its most direct action is through replying to requests from industry and research centres to find partners for research/ exploitation of research.

In 1982, Centres Régionaux d'Innovation et de Transfert de Technologies (CRITTS) were created. These are joint venture organizations with private and public (mainly regional) financial participation and the job of enhancing regional innovation-related networks between laboratories, firms and local governments. While these institutions are of great potential importance, the finance they command is marginal compared with the funds channelled to the core of the system (Chesnais 1993, 211).

7.   The Belgian innovation system

Belgium has a population of 11m, and is one of the most densely populated of the developed countries. Belgium consists of three regions: Brussels, Wallonia and Flanders and under the 1970 Reforms there are three Cultural (linguistic) Communities: Flanders (Flemish), Wallonia (French) and the German-speaking area on the border of Germany. Figure 2.1 shows the main regions in Belgium, the major universities and location of firms visited. In 1980, the regions were endowed with precisely defined powers and effective bodies. During 1988/9 the constitution was revised and special laws were adopted governing the Communities and Regions, their financial resources, the Region of Brussels-Capital (which did not have any administrative or decision making autonomy) and the Court of Arbitration. Since that period, the central state has approximately 60 per cent of the resources while the Communities and Regions receive about 40 per cent. National government decides macroeconomic policy – monetary, wages and incomes policy and tax policy, and social security. The individual regions are autonomous on cultural, social care, and since 1988 education (including the universities) has been administered through the Ministry of Education in each region. After 1993 when the Dehaene government

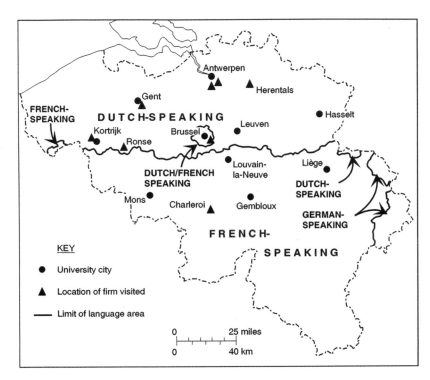

Figure 2.1 Map of Belgium showing main regions, major universities and location of sample

had succeeded in reforming the constitution, Belgium became a federal state with directly elected regional parliaments appointing their own regional governments. These reforms were followed by increasing 'centri-frugal tendencies' opening up the possibility of the complete disintegration of Belgium (Mommen 1994, 223). This was anticipated on the front page of *The Times* on 1 April 1992 with the headline 'Day of decision arrives for the future of Belgium'. This spoof story told of the absorption of Flanders by the Netherlands and of Wallonia by France.

Belgium is the exemplar of a country which has devolved respons- ibilities towards regional institutions, while policy issues are captured by international agencies (Swyngedouw 1994). Whilst regulation of important technological initiatives and programmes occurs at the regional level, international governance occurs through the EU and foreign-owned industrial organizations. Moulaert and Willekins (1987) traced the process of decentralization in post-war industrial policy and found that the real take-off for the Belgian economy came around 1960. From 1959 the incumbent Liberal-Christian-Democratic government began to display a selective economic policy. A system of investment subsidies was established to channel industrial restructuring into new growth sectors and to promote adaptation to the Common Market system. This instrument also was intended to attract international productive capital, particular US corporations seeking to benefit from the new European market. The government enhanced the good geographical position of Belgium within the EU and its reserves of semi-skilled and highly productive labour by public expenditures for infrastructure works, graduate education and scientific research.

The consequence of the open-door policy was that by 1987, overall about one-fifth of Belgian industry was under direct American control. This is reflected in the proportion of Belgian patents taken out in the US by foreign multinationals (19 per cent compared with a European average of 6 per cent) (Walker 1990, 13). Moreover, foreign trade in the late 1980s accounted for over 70 per cent of GNP (DTI 1990). Belgium and Luxembourg export more than any other OECD country. In 1991 over 75 per cent of Belgium's exports went to EU countries of which Germany (24 per cent) had become by far the most important trading partner (Mommen 1994, p. 222). Belgium has an average of 3 per cent of world trade.

In Flanders, a new 'industrial tissue' grew, largely controlled by foreign groups, while in Wallonia industrial activity remained (and still remains) based on traditional sectors, which have diversified to serve oligopolistic government markets (defence, the utilities, communications) and transport and financial services. In Flanders about 60 per cent of industry is controlled by foreign or mixed groups.

> Foreign capital's control of Belgian industry explains why the more progressive part of industry has only limited access to markets outside Western Europe; these are the protected markets of other daughters of transnationals which have made direct investments in Belgium (Moulaert and Willekins 1987, p. 316).

Indeed, recent research has shown that of 22 countries studied between 1971 and 1990, Belgium is the major beneficiary of research conducted in other countries, for example that undertaken in the USA, the biggest investor in R&D. The gains are mainly obtained through international trade (Coe and Helpman 1995, quoted in the *Economist*

18 March 1995, 112). The argument is that by importing from techno-logically advanced economies, countries acquire higher-tech inputs which make their own industries more efficient. Belgium's imports of research were worth nearly 90 per cent of GDP in 1990, compared to 11 per cent of the US's.

7.1   Support for innovation

Belgium has a less centralized system than either France or the UK. Federalization has meant an increasingly important role for the regions in supporting R&D. Examples of specific programmes illustrate this point.

Flanders

Flanders' technology policy which incorporates a science policy dates back to the beginning of the 1960s with the establishment of the Sociaal-Economische Raad Van Vlaanderen (SERV). This advises the Flemish regional government on socio-economic policy in Flanders, which includes economic, environmental, technology and science policies and the development of the regional infrastructure. It is orien-tated towards industrial uses, not university research. Its activities includes research on the social impact of new technologies.

DIRV is the overall policy of the Flemish government. This was the first important initiative of the first Flemish government. DIRV was an offensive policy. It was a reaction against the defensive national policy response to the crisis in the mid-1970s. It was to counteract the situa-tion where multinational companies in Belgium were doing well but Belgian industries were in crisis and firms were closing down. The inno-vation policy which started in 1981 stressed the development and appli-cation of high technology in firms. Universities were to be important and firms would be given research subsidies; universities and industry would work together. This was a push strategy in that basic research would be exploited: the Flemish region would have an important posi-tion in the international high-tech market and jobs would be created.

Public financing of innovation is an important component of economic development in Belgium. The Flanders' GIMV is a credit company, a semi-public investment company, financed from banks, the state and industry. Its board includes state and industry representa-tives. GIMV is instrumental to the formation and development, and rescue of Flanders' companies. Two notable examples are the electronic firms Barco and Mietec. Barco is still 51 per cent owned by GIMV; Mietec was originally 51 per cent owned by GIMV and 49 per cent by ITT and is now owned by Alcatel.

The Flemish government, in consultation with various interest groups, planned a number of initiatives aimed at promoting the development and application of new technologies in Flemish industry. At that time the concept of ASICs (application specific integrated circuits) was emerging. Initiatives resulted in the founding of such initiatives as INVOMEC, IMEC, MIETEC, INVENTIEVE SYSTEMEN. INVOMEC has the task of training engineers in microchips and despite having been established independently, it became based at IMEC (see Chapter 3). Unlike UK and French national laboratories IMEC has a very most cosmopolitan workforce. In 1993, foreigners constituted 15 per cent of the workforce, with 22 nationalities represented (Lawton Smith 1997).

The Flemish Foundation for Technology Assessment (STV) started in 1983. It also has an SME programme. The intention was to have a consistent plan to introduce micro-electronics by training SMEs to use micro-processors, and design chips. Three action programmes have been launched:

(i) micro-electronics – started 1982;
(ii) bio-technology – started 1989;
(iii) materials – started 1991.

IWT (Flemish Research and Technological Development Fund). IWT was established in January 1991. The regional government of Flanders decided to establish its own organization for stimulating research in industry and for stimulating co-operation between industry and university laboratories, concentrating all efforts into one organization. Previously there had been separate national organizations for industry and universities. In 1992, the budget of IWT for its own projects was about BF 150m a year. This included operations costs, salaries of staff and rent.

IWT assistance takes the form of a) grants for pre-competitive industrial research and b) interest-free loans for the development of prototypes or new manufacturing processes in Flemish industry. In the case of a), the maximum rate of award is 50 per cent of total project costs; in the case of b), the maximum award is 25 per cent grant equivalent (in other words the profit for the company, expressed as a percentage of the total project costs, obtained by taking the difference between the present value of all instalments and the present value of all repayments), where projects are successful and 40 per cent grant equivalent when projects meet with failure. A further 10 per cent is available in certain cases. These include SMEs, when the total eligible project costs exceed 50 per cent of the applicant firm's R&D expenditure for the current year; or is explicitly linked to a current EU project or programme.

IWT is a bridge between universities and industry, providing financial assistance to the interface between industry and universities, perhaps where industry tries to enter the EU schemes. Sometimes they are given help in establishing their own research programmes. For example the EU SME programme is a very important programme for companies in Flanders. Many SMEs in Flanders are in the field of materials, materials handling, transformation of materials. IWT provides support for entry to these programmes. A regional fund with a budget of BF 3 bn, FIOV, also exists to support research activities in industry. From this money about BF 800m was allocated directly to IWT in the early 1990s. The objective is to encourage firms to use the best technology available and to develop new technology.

Wallonia

In Wallonia, the Group of Technology Transfer (GTT) has been created by Walloon Region and Union Wallonne des Enterprises to promote technology transfer among SMEs. This operates under the aegis of the Ministry of Management of the Territory, Research, Technology and External Relations for the Walloon Region, under the 'Direction Generale des Technologies et de la Recherche'. GTT provides specific services to SMEs which are interested in acquiring or selling technologies or setting up technological collaborations with other companies. The services offered include technology auditing, patent advice and setting up technological collaboration.

8.   Concluding comments

This chapter began by identifying national and supra-national innovation systems in which measures to increase interaction between industry, universities and national laboratories are a general feature. The main features discussed include the major R&D performers, funding and technology transfer programmes. The identification of the main components of the UK, French and Belgian innovation systems show a number of distinctive features based on differences in philosophy.

1   The first difference between Belgium on the one hand, and France and the UK on the other, is that of devolution of innovation support to the regions. Belgium has a federal system. It uses both

direct investment in technology and regional funding mechanisms to assist in the process of creating locally constructed innovation filieres. In the UK and France programmes are co-ordinated at central government level. However, France unlike the UK allocates powers to target specific technology transfer objectives at the regional level, for example through ANVAR.

2 France still has a philosophy of supporting 'national champions' in electronics. This can be seen as part of the government's wider sectoral policy ambitions (Hancher and Moran 1989, 272). The UK, like Belgium has virtually no national champions left to support. French and Belgian heavy investment in the electronics sector is designed to ensure long-term development technological and production capabilities. Both the UK and France are characterized by heavy expenditure on defence R&D. Both systems have been criticized for their failure to translate this activity into commercial success in non-military markets.

3 A significant feature of the Belgian innovation system is the degree of openness to overseas firms, both in inward investment and by research interaction. Since the mid-1980s, the UK has followed this approach but France has continued to adopt a more protectionist stance to access to technology developed in its research institutions. This is illustrated by the case of LETI in Chapter 7.

4 Belgium's national system of innovation is dominated by industrial R&D to a greater extent than in other countries. The 1980s and early 1990s saw a deteriorating level of funding for public sector research. This balance might be expected to increase the influence of industry on the university sector to greater extent than in other countries where the higher education sector is protected by higher levels of funding. Likewise, the reduction in the UK on national support for innovation of both higher education and industry has increased pressure on industry to look for subsidized resources and higher education to look to increased industrial funding. This influence of the national innovation system can be seen in the level of participation by UK institutions in European programmes. This opening up of the funding for innovation does not necessarily mean a more equitable system of access to technology. Large firms, national governments and sectors are involved in negotiating for dominance in the system of allocation of resources.

In sum, the main differences between the three countries' innovation systems can be seen in terms of the approach to maintaining control over the national and regional space in which innovation takes place. Hence there is a difference between the kinds of openness which exists in Belgium to that in the UK. While Flanders in particular was attempting to underpin the power of the state to embed economic activities in the regions, the UK's failure to re-inforce the quality of the greater science base through under-investment and fragmentation of effort leaves the country more vulnerable to choices made by multinational companies about where they will locate their R&D activities. The centralized internal system in France on other hand, according to Chesnais (1993), has perpetuated the problems of 'lock-in' into less successful innovation and production systems.

3
Industry, University and National Laboratory Links

Introduction

Links between industry, universities and national laboratories are increasingly common, complex and spatially extensive. Recent studies have highlighted the growing tendency for firms to have some form of link, usually with a university but sometimes with a national laboratory. Most have dismissed the idea that innovation is a linear process and emphasize instead that it is an interactive and social process and that informal links between industry and the PSSB may be more important than formal interactions. Some research has focused on the geographical clustering of links between firms and universities while others have demonstrated the trend towards the internationalization of R&D activity. Yet another body of research has focused on the institutional context to identify how regulatory change has reconfigured relationships between universities, national laboratories and industry.

The purpose of this chapter is to summarize the main features of the organization of scientific and engineering research in the public sector in general in order to identify geographical, historical and political processes involved in shaping PSSB links with industry. It begins by examining geographical contexts in which industry and academic links take place. It then describes past and contemporary developments in relationships between industry, universities and national laboratories. This is followed by an examination of the characteristics of the PSSB in each country and then identifies the location of research relevant to the two sectors. The chapter concludes with some observations on the similarities and differences in the organization of the PSSB in each country and how they are likely to affect the extent and form of links with industry.

1. Interactions and space

1.1 Social factors

The development of informal interactions takes place in a variety of social, technological and geographical contexts. Social factors are becoming progressively more theoretically significant as explanations of the way in which industrial innovation occurs (De Bresson and Amesse 1991; Bergman *et al.* 1991; Sayer and Walker 1992, 115). Some authors, such as Faulkner and Senker (1995, 228) found that by far the most important flows of knowledge occur through informal interactions which take place outside any formal agreement. In their study, they found that links with universities were a significant contribution to innovative research, design and development (RD&D) through the education of scientists and engineers and through contributions to information generation and diffusion. Recruitment, while being basically a result of decisions by employers and employees, has the externality effect on contributing to the process of technology transfer. It is a means of facilitating the relocation of technological knowledge through personal networks which may have a regional focus (see Camagni 1994) but in academic circles are often spatially extensive. Membership of networks therefore facilitates technology transfer and creates a link between the cultures of different organizations (Van Dierdonck *et al.* 1990, 554; De Bresson and Amesse 1991).

In complex technologies, industry and university scientists and engineers share a common training having been through the same or equivalent education system. Moreover, their professional interests overlap. As Nelson (1989, 233) has pointed out, corporate R&D is an important source of public generic knowledge; in some fields, professors invent; and the lines between what corporations do and what goes on in universities are often blurred, not sharp (Nelson 1989, 233). As a result, both industrial and academic researchers may belong to 'technological communities'. These are defined as 'a group of scientists and engineers, who are working towards an interrelated set of technological problems and who may be organizationally and geographically dispersed but who nevertheless communicate with each other' (Debackere and Rappa 1994, 356). Debackere and Rappa make the distinction between a technological community and a scientific community, arguing that there may be distinct demarcation between those scientists and engineers who network extensively and whose links are more focused on particular types of partner. Whereas a *technological community* is interdisciplinary, a *scientific community* is often demarcated by highly specific disciplinary boundaries. The authors suggest that increasing intensity of university-industry R&D linkages may foster more 'communalism in industrial R&D, and may force industrial R&D organizations to become more active participants in activities that are relevant to their core competencies'. Moreover, technological communities are sustained by the movement of scientists and engineers between other sections of the science base, either by changing jobs, or temporary residence such as secondment.

1.2 Spatial construction of networks

In the mid 1990s universities contribute to a wide range of economic development strategies. These include addressing lagging productivity growth, sparking regional development, and improving the competitive position of nations' industries in world markets. The spirit of techno-nationalism has cast universities as an instrument of national R&D policy, assigning them a central role in generating knowledge and transferring it successfully to the domestic sphere of industrial application (David *et al.* 1994, 14). The rationale for their involvement is based on the belief that universities' competencies can be developed and employed to gain national economic advantage.

The importance of proximity is gathering momentum as an analytical theme in the exploration of the nature of the innovation process. Explanations for the degree of geographical concentration of innovators have been put forward by Breschi and Malerba (1997, 141). They suggest the following:

1 Innovators are geographically concentrated when there are conditions of high opportunity, high appropriability and high firm cumulativeness (in this case spatial concentration overlaps with sectoral concentration) a relevant source of scientific knowledge available in a specific location, or a knowledge base characterized by tacitness, complexity and systemic features.
2 Innovators are geographically dispersed when there is low opportunity, low appropriability, and low firm cumulativeness, or a knowledge base is relatively simple and codified.

The role of universities and national laboratories as a component of the clustering of innovative activity arises because they are 'like agglomerations of industrial firms, repositories of localised sets of specific competencies that have been built up through active participation in research endeavours' (David *et al.* 1994, 14). Linkages between innovative firms and universities, and in some countries national laboratories, are shown to be clustered. The ten 'islands' which concentrate innovation activity in the European Union are notable for the local concentrations of science-technology interactions and university-industry collaboration (EC 1994, 203). In the USA, a clear pattern of the spatial range of interaction was found by Anselin *et al.* 1997. They found that spillovers of university research on innovation extended over a range of 75 miles from the innovating Metropolitan Statistics Area (MSA), and over a range of 50 miles with respect to private R&D.

Clustering around universities can be related to different stages in the development of an industry. There is evidence of a tendency towards spatial concentration of R&D activities because proximity is important when technology changes rapidly (Lundvall 1988). For example biotechnology is a university/research institute based industry (Kenney 1986, Faulkner and Senker 1995), and new firms and the research laboratories of pharmaceutical companies are found to be clustered around major centres of expertise (Arthur Andersen 1997).

Although examples such as the biotechnology industry can be found to support the argument that proximity and technological change are related, Storper (1997, 16) has pointed out, more consideration needs to be given to what it is about the kind of knowledge or its transfer to producers that makes R&D presence effective sometimes and not effective in other times and places. For example, while universities were important in the growth of high-tech industry in Silicon Valley and Massachusetts, they were not strongly present in the case of aircraft in Los Angeles in the 1920s and 1930s. He finds explanations of what set off and sustained these linkages unconvincing when they are taken to be a universal logic of new technology-based infant industry development. Moreover, the regional politics approach which holds that 'regional coalitions secure resources that push for the transfer of high technology resources' also fall short of a complete explanation. Storper favours the evolutionary school which permits the identification of the intangible aspect or territorial or regional economy that underlies innovative flexible agglomerations of both high and low tech variety (Storper 1997, 18). A key concept in this approach is that of 'technological spillovers' which derive from supportive specialized public and private services. These are described as 'knowing how to do one thing is frequently consequent upon knowing how to do another, or key to doing certain other things' (Romer 1990 in Storper 1995, 204). In the new theory of endogenous economic growth which Storper currently finds attractive, the accumulation of knowledge and its spillover into productive capacity through technological change results from the territorialization of technological spillovers and their untraded inter-dependencies (Storper 1995, 206).

Industry-academic links provide technological spillovers through a number of different kinds of interaction. For example training, labour mobility and spin-off activity within locality both increase flows of knowledge and develop individuals' skills in interpreting knowledge. Training of industrial staff both in the course of collabo-ration and in-situ type of activities such as industrial residency in university laboratories encompasses understanding and translating new knowledge. Likewise, collaboration with industry can help acad-emics recognize the potential commercial value of their research and communicate this to their industrial colleagues. Local recruitment and academic entrepreneurship can have similar spillover effects. Each of these forms of spillovers then contributes to the growth of localized networks which intersect with more spatially extensive personal networks.

Another important influence on the development of spatial concen-trations of expertise is that of locational decisions on the R&D activ-ities of transnational firms within and between countries. Concentrations of elite firms occur at different places at different times, shifting the locus of innovation networks and the spatial organization of innova-tion filiere. These both are influenced by the supply of skills and re-inforce existing characteristics of local scientific and technical labour markets. Local R&D activities are themselves part of a super-regional, often world-wide manufacturing network co-ordinated by a few core research centres (Moulaert and Swyngedouw 1992, 58). However, the view that the multinational corporation functions as an integrated, spatially extensive system of flows implies that local multiplier effects (and local innovation linkages) may be severely circumscribed. Decentralized branch plants are indifferent to external economies within local/national innovation systems (Schoenberger 1989, 123–4). Moreover, returning to the work of Breschi and Malerba (1997, 142) it may be that there is a significant spatial dimension to many learning activities which confine them within national boundaries. This is when innovation involves the integration of codified, specialized, and system specific pieces of knowledge for example in the automobile industry.

From the perspective of individual nation-states, there are advan-tages and disadvantages of domestically owned firms locating R&D

outside their borders and of hosting the R&D activities of foreign firms. On the one hand there are advantages in hosting the largest amount possible of skilled activities. Foreign investment may either: reinforce and develop research undertaken in the science base (depending on the terms of exploitation), where for example, a foreign firm takes a longer-term view than a domestic company or no domestic company is forthcoming, or (also depending on the terms of exploitation) relocate the benefits of research to their home or another country. In the latter scenario, externality effects of research in the form of technology as a public good are lost to the domestic system.

Recent evidence suggests a possibility of branch plants, not noted for engagement with the PSSB, developing a greater degree of technological embeddedness locally (Hudson 1995, 4.5). Hudson suggests that these would be factories in which there are some R&D functions, perhaps with more skilled jobs, with the plant rooted for a longer period by virtue of its relations with the surrounding area. In this case, embeddedness would form part of its competitive advantage. Similarly, Turok (1993a, 402) argues that the process of vertical disintegration of large corporations and decentralization of decision-making demands closer and more collaborative relationships between individual plants and suppliers and distributors within the value chain. This restructuring of responsibilities encourages stronger geographical clustering to minimize transaction and transport costs and facilitates high-level exchange of technical ideas, market awareness and corporate plans.

Clustering may be strongly influenced by regulatory intervention as well as by technological change and organizational strategies. For example, the local impact of universities and national laboratories in some places is conditioned by nationally determined factors such as funding of research, accountability in terms of politically determined agenda and their position within national innovation systems. Engagement in the local economy in some countries is a function of public policy. For example, in Munich each university and technical college has an agency responsible for technology transfer within the high-tech region of Munich as well as to all other areas (Sternberg 1996, 12). The Inter-university Micro-electronics Centre (IMEC) in Flanders is directly funded by the Flanders government to help SMEs, provide training for industry, schools and colleges, and attract inward investment (Lawton Smith 1995). On the other hand, institutions acting independently of a national policy framework can be major contributors to local economic development through links with local industry. For example, some have made a deliberate commitment to make links with industry a key element in the University's long-term development strategy, as in the case of Warwick (Segal Quince Wicksteed, 1988, 17–22). In the USA some universities have 'eagerly taken up the challenge of regional economic development' (Etzkowitz and Stevens 1995, 23).

In some countries universities' technology transfer role is incorporated into a property-led strategy. For example, in France, innovation and technology parks related to dominant industries in regions, include government research institutes and big firms' R&D departments. Technology parks are instruments of national innovation policy. They 'are expected to contribute to industrial restructuring and to manage regional problems by contributing to strategies for regional innovation that are based on techno-scientific information and the adoption of new tech in regional economies' (Hilpert and Ruffieux 1991, 65). Belgium has also adopted a science park policy as a means of encouraging technology transfer at the local level (Van Dierdonck 1991).

Spatial concentrations of innovative activity in the case study countries can primarily be found in the South East of the UK, in Paris and the South-East of France, and in Flanders in Belgium. In each of these cases, a primary cause of concentrations of innovative activity has been the location of government funded research centres. The South East of the UK has most of the country's industrial R&D units (Buswell, Easterbrook and Morpeth 1985), MoD procurement offices and officers, and government research laboratories (Heim 1988). It has a share of significant innovations well above its manufacturing employment, the highest incidence in the UK for product innovation, and is most likely to benefit from the transfer of foreign technology into the UK (Goddard *et al.* 1986, 147). Paris has a quarter of all French R&D. A second centre is in the Rhone-Alpes region (Beckouche 1991; Hilpert and Ruffieux 1991). Grenoble for example, has a remarkable concentration of education and training functions; it has three universities, the Institut National Polytechnique de Grenoble (INPG) plus *grandes ecoles* and technical colleges, (Dunford 1989, 86). A series of government decisions on decentralization of public sector institutions dating back to the 1950s reinforced and extended the city's scientific and engineering base. In 1956 CEA moved one of its five civil centres to Grenoble, which resulted in the creation of the Centre d'Etudes Nucleaires et de Grenoble (CENG) (Dunford 1989, 86). In Belgium, recent policies have perpetuated Flanders' position as the region with most public and private research activity.

1.3   Internationalization of R&D

An increasing trend is the internationalization of R&D (Howells 1990). Technological knowledge and capability, and in some industries such as pharmaceutical and electronics, institutions are international rather than national (Nelson 1993, 75). Archibugi and Michie (1995, 121–40) find that international access tends to occur in order to acquire knowledge, to acquire know-how which is lacking at home and not to replicate research and innovations in sectors where the home country is already strong. This is linked to international technological specialization. International operations of large firms are increasingly exploiting and developing this diversity. However, it is possible to over-state the generality of this trend. Archibugi and Michie find:

1   The story of globalized R&D is the story of a fairly small number of very large firms carrying out research in a small number of leading industrialized countries.
2   International generation of technology has been to date a peculiarly intra-European phenomenon, and has been a character of European *regionalization* rather than globalization.
3   US and Japanese firms have not pursued the global generation of technology to any significant extent.

The authors hypothesize that global exploitation of technology is a consequence rather than a cause of the increase in international trade. This finding is consistent with research in the 1980s when it was argued that international appropriation of public research had lagged behind global production patterns (see Howells 1990).

   In sum, proximity may be important in some cases, but the reality is that many 'knowledge-intensive firms' operate within networks of high level contacts with a number of institutions which are not local (Keeble *et al.* 1997). Two questions which have not been dealt with adequately in the literature are:

1   How are places, i.e. the combination of firms and institutions, positioned within wider innovation systems (Malecki 1990, Gertler 1990, Swyngedouw 1995)?
2   How can the impact of such public-private partnerships on localized structural change be evaluated (Ettlinger 1994, 144)?

2. The development of university, national laboratory and industry links in Europe

University and national laboratory departments differ in the roles they play in the technological life cycles of industry, and their position in the spectrum of 'pure' to 'applied' science and engineering. Their functions are derived from their historically determined position within the hierarchy or structure of the each national innovation system. These points are developed by an examination of the historical context to industry, university and national laboratory links in the three countries.

2.1 Universities

As Charles and Howells (1992, 10–12) record, industry and academic links are not a recent phenomenon. Industry and academic links in Europe and in the USA date back to at least the nineteenth century. Even then they were not without problems. Charles and Howells found that the same tensions and motivations which existed a hundred years ago still exist today. Then as now academics were concerned that collaboration with industry was against the central ethics of universities and had undesired side-effects. These centred on familiar problems of restrictions and distortions on the free flow of information and materials within the academic community. Links between industry and universities in the first half of the twentieth century were shaped dramatically by wars, and the threat of wars. The relationship between defence research and academic research, which is now pervasive in the UK university sector (Evans *et al.* 1991), has its origins in the need to mobilize scientific effort for defence. The UK government, like those in countries such as those of the US and Japan, intervened in the process of innovation bringing industry, academics and government scientists together in united efforts. An example of the strategic role played by universities was that in the 1930s when physics laboratories in Cambridge, Oxford and Birmingham universities were mobilized to work on radar, and other defence-related activity.

By the 1970s, there were the beginnings of widespread formal mechanisms to exploit academic research in Europe. In Belgium, Leuven R&D, a university technology exploitation company which developed out of KUL, was founded in 1972 as a non-profit organization. It was the first such unit to start in Western Europe. However, it was not until the mid-1980s that such institutional arrangements became common. In the late 1970s and early 1980s, European, particularly UK and

French academic literature, recorded that progress between establishing productive links had been slow (see for example ACARD 1983; Chanaron 1989, Brunat and Reverdy 1989). In the UK, both industry and academia were blamed for conservative attitudes which inhibited interaction. Industry and academic links were haphazard and *ad hoc* (Howells 1986).

By the mid-1980s the different aspirations of industry universities and industry began to converge and a variety of means were developed to explore 'mutual benefit'. A reflection of increasing interaction was the establishment in 1986 of the journal *Industry & Higher Education*. This is dedicated to 'improving the interface between business and higher education'. It contains articles from the academic, policy and industrial communities in Europe, North America and Asia. Such mediums both reflect and are part of the changing culture, vocabulary and belief systems in universities and of industry's perception of the value of universities.

By the 1990s, relationships between universities and industry had become much more structured than in earlier decades. The combination of the increasing priority given to the funding of industrially relevant research in universities and national laboratories and the growth of mechanisms designed to encourage collaboration (see below) has meant that research undertaken on behalf of, or in conjunction with, or with the views of users (industry, policy makers and so on) in mind, had become a norm in British, Belgian and to a lesser extent in French universities, and in British and French national laboratories. The interests of universities and those of industry had become 'compatible' even if the closer links with industry were universally welcomed by the academic community (David *et al.* 1994, 19). Studies found that academics described the benefits of working with industry as including exposing students and staff to real life industrial problems, improving the content of lecture series and obtaining funding for research that would otherwise not have been undertaken (Lawton Smith 1990; Bonaccorsi and Piccaluga 1994). Most universities now have a well developed range of formal mechanisms designed to encourage and facilitate linkages. In the UK these include such as Industrial Liaison Offices (ILOs), and university-owned know-how exploitation companies.

The wide range of interaction plus problems of establishing objective criteria for evaluation of the outcome of closer links mean that the benefits of links to industry on the one hand, and universities on the other, are unquantifiable. Assessment is both an issue of data availability and of perception of what the outcomes should be within a temporal framework. Qualitative assessment takes a variety of forms. Bonaccorsi and Piccaluga (1994, 242–3) distinguish between three dimensions of performance, knowledge generation, knowledge transmission, and knowledge dissemination. Their view is that evaluation of performance of industry and academic links should take account of the generation of new opportunities and goals, for example unanticipated learning, discovery of new co-operation fields and change of action plans and productivity of basic research activities.

The measurement of outcomes is inevitably inexact. For example, patents as a measure of the output of universities are largely unrecorded, because of problems of lack of records from universities and that patents are an imprecise indicator of innovation for various reasons (Webster 1991, 196). The value of training provided is another criterion of how effective universities are in stimulating industrial innovation. Many fields of academic science are important to technical change because of the training they provide young engineers and scientists who go into industry (Lawton Smith 1995). Perhaps the most significant finding on the profitability of funding university research is that of Link and Rees (1990). Their study of 209 companies in five sectors found that firms with university links gained a 34.5 per cent return on their R&D expenditure compared with only 13.2 per cent for those without such links.

In sum the 'usefulness' of universities is now measured not just in terms of 'socially desirable level of knowledge' (Rosenberg 1990, 167), but also as generators of information which can be commodified and monopolized (Lawton Smith 1991, 405). Universities' primary and unique function may still be that of a 'node in the open knowledge generating network' as their (David *et al.* 1994, 19). For these authors, the issue is not whether such involvements are mutually beneficial to the participants or the public at large. Rather they take the issue to be 'the terms on which such involvement can be established' and whether they will or will not be compatible the university's ability to function as an open knowledge-network node'.

Access to and control of university research is increasingly determined by the ability to pay while the commercial value of certain fields of science is expanding, for example biotechnology and materials science. This trend and the political priorities of the Conservative government in the UK (1979–97) changed the nature of accountability in the university sector as industry increasingly became involved in the planning and funding or research programmes. Whiston (1992,

173) argues that where a system is deferential to government, to industry or the wider corporate framework (as it is the UK), then an inevitable shaping of the research culture is implied. Moreover, subordination of research activity to the needs of industry requires new kinds of (industrially relevant) competencies from academics (Gibbons 1992, 99). In turn changes in prevailing expectations affects academics' actual and perceived bargaining positions in relationships with industry.

Yet in spite of assumptions made in much of the literature, direct technology transfer between universities and industry may not be the most significant link in longer-term economic development. Some research suggests that it is the impact on the local labour market that is the prime contribution to regionally based innovation activities (see Lawton Smith and De Bernady 1998). Moreover, it is a current concern whether academic research and commercial activities are compatible in the long term. Feller (1990, 343) argues that

> The existing tracks upon which academic research flows to the market are likely to become blocked if not broken apart as universities limit existing flows of information in order to divert faculty findings to specific firms. The consequence is lower rates of technological change.

This is a similar line of thought to that expressed by Etzowitz and Stevens (1995, 24). They argue that a focus on technology transfer actually produces a barrier between the university and industry by displacing the free flow of knowledge which takes place through publication and informal discussion. Technological collaboration may reduce competition and hence the pace and degree of innovation, by damaging informal networks on which firms rely (Macdonald and Williams 1994, 145). Policies towards exploitation also ignore historical evidence that technological innovation may result from the exploitation of research results that are decades old and often in seemingly unrelated fields. The long-term consequences of close interaction is a matter of debate. This is reflected in the academic literature, from public statements by academics and by organizations such as the UK university based pressure group 'Save British Science'.

2.2   National laboratories

The foundation for national research institutions in Europe were laid at the end of the nineteenth and early twentieth centuries. For example the UK's National Physical Laboratory was formed in 1900, and France's National Aeronautical School was founded in Toulouse in 1909. During the second world war, UK private laboratories were brought into the public sector to enforce collaboration within the public sector, bringing all defence R&D under single management (Charles and Howells 1992 11/12).

Post-war national laboratories were set up by European and North American governments to promote large scale R&D programmes in strategic areas such as energy, defence and space (Heim 1988, 376). However, it was the post-war commitment to nuclear energy, to which nearly every OECD member in the post-war period devoted substantial R&D resources which 'marked the beginnings of big science and big technology' (OECD 1989b, 21). Space became the other spectacular growth area of government funded research in the post-war period. These two new developments led to an expansion in the government research sector with the creation of new laboratories geared to particular needs, for example in the UK the National Mechanical Engineering Laboratory (later the National Engineering Laboratory (NEL) was set up 1946. In nuclear energy, France established the CEA in 1945, Belgium, the Studiecentrum Voor Kernenergie (SCK) in 1952 and the UK, the United Kingdom Atomic Energy Authority (UKAEA) in 1954.

From the mid-1960s, most OECD countries began to reassess the role of national laboratories. During the period 1960–80 major programmes (defence, nuclear energy, telecommunications and civil aircraft) consumed major national resources. In the UK, for example, this amounted to two thirds of UK's R&D outlays (Ergas 1993, 6). A re-ordering of priorities led to a scaling down of research effort and diversification of activities. All three countries in this study began to introduce measures to increase greater utilization and exploitation of national scientific and technological resources through their use in industry. For example, the 1965 Science and Technology Act allowed the UKAEA to undertake non-nuclear projects. The UKAEA was given Trading Fund status in April 1986. This status meant that from then on 100 per cent of income had to be earned by contracts from customers. In France, LETI was established in 1967 by the French Atomic Energy Authority (CEA) with the objective of the valorization of research initially undertaken for the nuclear energy programme. CEA Industrie was split off from CEA in the early 1970s and is the holding company for the French Atomic Energy Commission. Its purpose is to apply the CEA's research and development to high technology sectors.

For national laboratories, the last decade in Europe and in North America has seen radical reassessment and redefinition of the role of national laboratories (Lawton Smith 1995; Branscomb 1993) and repositioning within the national innovation system. This is due to the ending of the cold war, maturity of nuclear energy technology and the associated problems of waste disposal and decommissioning of old reactors and power stations. These have presented common problems on how to adjust their functions to meet the needs of the 1990s and beyond. The major differences between the UK on one hand, and France and Belgium on the other, are in the current solutions to how to maximize the economic benefits of this long-term state investment through increased interaction with industry.

The UK, far more so than in France and Belgium, moved away from national laboratories' 'mission orientated' programmes (Ergas 1993). The balance of research has moved away from the longer-term theoretical research (academic model) towards short-term applied research (consultancy model). France too has moved towards commercialization, although it has not adopted a wholesale strategy of privatization. An important difference between the UK, and France and Belgium, is that longer-term commitment to large-scale programmes such as nuclear energy, allows the research centres to have a more strategic role in regional development. The UK's laboratories increasingly have short-term commercial targets and are not funded to meet wider socio-economic objectives. However, they like those in France and Belgium have had some functions in common with universities. These include student supervision, staff lecturing in universities and joint research. While these persist in France and Belgium, they have declined sharply in the UK (Lawton Smith 1995). This has had the effect of distancing them from academic networks to which they previously belonged. National laboratories relevant to this study are shown in Table 3.1.

3.   Universities and national laboratories in the UK, France and Belgium

The major difference within the three national systems of innovation are the respective functions of universities and national laboratories. Universities in the UK and Belgium are a central component of the innovation system, while in France, national laboratories and CNRS research groups undertake the majority of industrially relevant research in the public sector.

Table 3.1 National laboratories in the UK, France and Belgium showing research specializations, funding and status

Laboratory	Technological field	Ownership	Status
UK			
Atomic Energy Authority, Harwell Laboratory	Originally nuclear energy, in recent years non-nuclear areas: industrial technology, petroleum services, environment and energy, safety and reliability, and engineering	DTI	Privatization of AEA Technology, March 1996 UKAEA remains in public ownership
National Engineering Laboratory (NEL)	Flow measurement, energy and the environment	DTI	Privatized, October 1995
Defence Research Agency (DRA) Malvern (since known as Defence Evaluation and Research Agency, DERA)	Electronics, command information systems	MoD	Became 'Next Steps Agency' 1991
France			
Centre d'Etudes et de Recherches de Toulouse (CERT)	Fluid dynamics	Defence Ministry	Operates under defence ministry but only partially owned by the state. Constituent institution of Office National d'Etudes et de Recherches Aerospatiales (ONERA), the French aeronautics organization
Laboratoire D'Electronique, De Technologie et D'Instrumentation (LETI)	Micro-electronics	Defence/ Industry Ministry	Part of the French Atomic Energy Commission (CEA)

Table 3.1 Continued

Laboratory	Technological field	Ownership	Status
Belgium			
Inter-university Microelectronics Centre (IMEC)	Micro-electronics	60 per cent Flanders government	Flanders government plans to decrease share of support from Flanders
Studiecentrum Voor Kernenergie (SCK)	Radiation protection, fuel research, reactor materials, dismantlement, waste and disposal	Ministry of Economic Affairs	National research laboratory
Vlaamse InstellingVoor Technologisch Onderzoek (VITO)	Energy, environmental and biotechnologies	Flanders government	Spin-out from SCK
Von Karman Institute (VKI)	Fluid dynamics	NATO	Former military research establishment. Now non-profit scientific association

3.1   UK: the university sector

The defining characteristics of the UK higher education system are (a) its central position within the national system of innovation but relatively low share of public funding for university research, (b) the relative autonomy of institutions to decide their technology transfer arrangements and (c) the high level of defence research activity.

Central position/low share of public funding

In the early 1990s, about 90 per cent of the research in the higher education sector was done in the universities, this sector being the biggest performer of basic and strategic research in the UK. The polytechnics and colleges accounted for the remaining 10 per cent of the HE sector. The universities' work was mainly supported from their own general income together with 'third party' funds, either in the form of direct grants from research councils for specific projects or for fellowships, or from the EU, or as contracts from industry or from government departments. Less research was undertaken in the polytechnic sector and colleges of higher education (Atkinson *et al.* 1990, volume II, 13).

In spite of the importance of university research in the national innovation system, the UK has one of the smallest proportions of public funding for university research of OECD countries. Over the decade 1981–90, there was considerable growth in funds from charities, industry and overseas sources which in 1990 contributed about 10 per cent, 8 per cent and 5 per cent respectively. Government funding fell from 81 per cent in 1981 to 72 per cent in 1990; in the USA the proportion was 86 per cent in 1990, in Germany 92 per cent and France 93 per cent (SBS 1994, 5). The most dramatic change was the decrease in more than 15 points of the Exchequer Grants' share in university funding over the period 1989–93 (David *et al.* 1995, 39–41). This was a result of both the policy of funding a higher share of university research directly through specified projects and the decision to link part of teaching money to the number of students, via the increase in fees. EU funding although low, more or less doubled over the period. The research councils' share of funds, after a slight decrease over two years, increased in the last year. This was due mainly to the new regulations would become responsible for meeting all the costs of projects, which includes use of research facilities such as libraries. The existence of a large number of universities means that research is geographically fragmented and there is extensive competition for research funding between the many universities and more recently with the national laboratories.

Another trend identified by David *et al.* is that both research councils and EU have focused their top support on a similar group of institutions – the largest (such as Oxford and Edinburgh) and/or the research orientation, such as Imperial College. The authors conducted a study of UK participation of the EU BRITE-EURAM I and II programmes. They found that ex-polytechnics and other higher education institutions rarely succeed in getting funds from the EU. One of the effects of EU funding may, therefore, be to further concentrate research funding within the UK. Table 3.2 illustrates sources of funds in UK departments.

The problems of decreasing levels of funding have caused considerable dissatisfaction in the university sector. It was not usual in the mid-1990s to come across criticism of the UK system of innovation expressed in academic literature and in the media. For example, in a

86

Table 3.2 Sources of funds of UK universities

Year	EC	Other specific	Research councils	Govern. bodies	Exch. grants	Fees
1989–90	1.1 per cent	22.0 per cent	6.5 per cent	3.0 per cent	48.4 per cent	13.8 per cent
1990–1	1.3 per cent	22.9 per cent	6.4 per cent	3.3 per cent	39.8 per cent	20.7 per cent
1991–2	1.6 per cent	22.4 per cent	6.0 per cent	3.1 per cent	35.5 per cent	25.0 per cent
1992–3	1.9 per cent	23.2 per cent	7.1 per cent	3.1 per cent	33.3 per cent	25.3 per cent

Source: Merit elaboration of Universities' Statistical Records data, in P. A. David, A. Guena and W. E. Steinmueller (1995)

recent newspaper article, Professor Michael Harlowe, Pro-vice Chancellor of Essex University wrote:

> The cumulative process of running down our research and training base affects our ability to compete with the best foreign institutions. It is the big scientific projects that are affected. We are losing out in whole areas like biotechnology and telecommunications (*Observer*, 25 February 1996).

In the late 1980s the UK began a series of changes to the structure of the higher education system. David *et al.* (1995, 36) summarized the major changes. They identify the turning point in the process of change of the funding system as the Education Reform Act (1988). The Act created two new funding Agencies, the Universities Funding Council (UFC) and the Polytechnic and Colleges Funding Council (PCFC). These modified the 'logic' of higher education funding. The two agencies were created as 'buyers of academic services'. In 1993, the UFC and PCFC were merged into a single Higher Education Funding Council (HEFC) with separate agencies for England, Scotland and Wales. At the same time the higher education sector was transformed with 39 polytechnics and colleges being granted university status. Other institutions subsequently became universities. Under the old system in 1993, there were 74 different universities, 52, if London University is counted as a single institution rather than counting its 23 component institutions. By 1995 the complement of universities with undergraduates was 96 (*The Times* 1995). In addition there are universities such as Cranfield, which are graduate only teaching institutions and which also undertake research.

The White Paper published in May 1991, *Higher Education: a New Framework*, set the agenda within the higher-education system. David *et al.* (1995, 37) identified three central tenets:

1 As well as defining the new structure, it favoured competition among all of the university sector.
2 It endorsed and reinforced the nation of dual-support approach. Universities should receive public funds for research from both the national HEFC and from research councils for specific projects.
3 It introduced the notion of competition among institutions and the dual system, selectivity on the basis of assessment of research quality and the subdivision of the block grant in teaching and research are the principles that inform the new structure for public funding.

Relative autonomy of decision making

Autonomy of decision making over rules governing technology transfer arrangements dates back to May 1985 when the British Technology
Group's right of first refusal of inventions arising from publicly funded
research ended. Since then, Higher Education Institutes (HEIs) have
direct responsibility for ensuring that inventions from publicly funded
research are exploited but it is up to individual institutions to decide
on arrangements. These vary considerably in the terms-of-trade which
are set and therefore influence access and openness of academic
research to external organizations. For example, the University of
Oxford founded ISIS Innovation Ltd, a wholly owned company of the
University of Oxford in 1988. Its function 'is to ensure that the results
of research are exploited wherever possible and bring reward to the
University and to the inventors'.

High level of defence research activity

A characteristic of the UK higher education system is the extent to
which government funding of defence-related research projects controls the direction of research. In 1991/2 the Ministry of Defence was
spending £65 million to fund 851 research projects in universities and
polytechnics to fill a gap or complement scientific effort in the MoD's
own research establishments. These were spread around virtually every
university in the country as well as up to 15 polytechnics (Evans *et al.*
1991, 16). The MoD has been increasing its funding of science in
higher education. Between 1984 and 1991, annual expenditure tripled.
The majority was for long-term basic research. MoD expenditure on
research in 1991–2 was £18 million, twice as much as on development
(£9 million). The formation of the Defence Research Agency in 1991
was expected to increase expenditure (Evans *et al.* 1991, 18).

3.2   UK national laboratories

In the 1990s a major change in the organization of the science base
was brought about by changes to the status of the national laboratories. This diminished their research and expanded their commercial
function. The most important of the new measures is the creation of
Executive Agencies under the 'Next Steps' initiative which began in
February 1988. The initiative's objective was to bring improvements in
the quality and efficiency of government services. Within such agencies, responsibility is delegated to a Chief Executive within a framework of policy objectives and resources set by the responsible Minister
in consultation with the Treasury (POST 1993, 1). By the end of 1992,
76 Executive Agencies, employing 290,000 civil servants had been

created. NEL became the ninth DTI Executive Agency in October 1990. The Defence Research Agency (DRA) was established as a Next Steps Agency in 1991, and has operated as a Trading Fund from 1 March 1993. It has remained in the ownership of the Secretary of State for Defence. Most government research establishments (national laboratories) are now run as agencies. For some non-defence laboratories, for example NEL, the National Physical Laboratory, the National Weights and Measures Laboratory and the Laboratory of the Government Chemist, the conversion to agency status was in preparation for privatization. NEL was privatized in October 1995.

The 1993 White Paper, 'Realising our Potential: a Strategy for Science, Engineering and Technology' set out a range of policies and initiatives designed to improve the nation's competitiveness and quality of life by maintaining the excellence of its science, engineering and technology. A major theme was to promote a strong science base and ensure efficiency and effectiveness of government research and development. Following the White Paper, the government commissioned the Multi-Departmental Scrutiny of Public Sector Research Establishments (MDSPSRE) which reported in 1994. The review included 50 establishments parented by government departments and related laboratories within the research council system in England Wales and Scotland but excluded defence establishments ((MDSPSRE 1994 13/14). These PSRE comprise a substantial part of the UK's civil and technology base (31,000 staff; budget £1.3bn). Although comprising a substantial proportion of such bodies, the coverage was not comprehensive on scientific grounds (for instance only five of the 46 MRC institutes or units were included).

Two important questions not raised by this review were:

(i) whether the expansion of commercial activities would make better use of existing resources. This would depend on the degree to which the 'market is free of distortions' (POST 1993) and whether this was a sensible expectation given that there are other organizations such as PERA able to provide similar services;

(ii) whether it would be possible for these institutions to act as providers of core services and expert information to government departments while acting in a fully commercial way when tendering for contracts in the market (Webster 1994, 147).

3.3   France: the university sector

This description of the French higher education and research systems is taken from Atkinson *et al.* (1990). They describe the French higher education system as comprising the universities and the *grandes ecoles*. Universities are funded by the Ministry of Education, but little money for research comes from this ministry. Unlike in the UK, universities in France play a relatively weak role in research. The *grandes ecoles* traditionally did no research at all, and thus their graduates, filling most of the top jobs in France, had little exposure to the research environment. Instead, basic and strategic research is primarily carried out under the province of MRT's national centre for science research CNRS. Indeed most university academics who do research do so in a CNRS laboratory. In recent years, France has moved to strengthen research in universities. Each university now has a research plan and a four-year contract from the ministry of national education, (MEN) and the top *ecoles* are increasingly engaged in research.

3.4   France: the national laboratories

In France much applied research, including research in support of industry, is carried out in government research establishments or agencies including those for atomic energy (CEA) for space (CNES) and for telecommunications research (CNET) (Atkinson *et al.* 1990, 4). Research in public sector laboratories is more generously funded and differently organized in France than in the UK. Public sector establishments perform about half of France's military R&D with a spend of FF11.8bn in 1989. These establishments include those of the CEA, of the Office National d'Etudes et de Recherches Aerospatiales (ONERA), the national office of aerospace studies and research, of the Ecole Polytechnique and of numerous university and public laboratories (Atkinson *et al.* 1990, 29). The CEA is the single largest French government R&D centre (Chesnais 1993, 207), but is smaller overall than the Centre National de la Recherché Scientifique (CNRS), which is the largest basic research agency in Europe. By the end of the 1960s there were some 30,000 employees in the CEA. The CEA decreased in size in the late 1980s and early 1990s to bring total employment down to about 18,000 by 1993/4. The population was predicted to stabilize at around 17,000 in 1998 with 12,000 in the civil programme and 5000 in defence. In 1992, military programmes accounted for nearly half (47.2 per cent) of CEA expenditures, compared to 6.8 per cent on advanced technologies (industrial technologies). In the late 1980s, CNRS employed 26,400 in 1338 research laboratories (CNRS Annual Report 1990). Relations between CNRS and CEA are organized in particular areas – such as the establishment of a large institute in Grenoble acting as the focus of the CEA Protein 2000 programme and the CNRS Imabio programme.

Commercial technology transfer activities from CEA advanced earlier than in the UK. LETI is the main part of the 'advanced technology' programme of CEA. It was formed in 1967 with the objective of the valorization (commercialization) of the work which was initially done for nuclear energy. This is achieved primarily by developing new technological processes in microelectronics which would be licensed by industry. The period 1970–6 saw the privatization of all CEA activities relevant to industry, through the formation of CEA Industrie. This is a financial holding for nearly 200 companies in a variety of sectors. This period represented a break from the first phase of the CEA's operation which was purely research directed to energy and defence. The current phase is the most difficult for CEA as it is redefining its research objectives and strategy in defence and nuclear energy (personal interview 1994).

Two major changes occurred in LETI in the late 1980s and early 1990s. The first was that research funds for both longer-term programmes and for applied research from industry had become harder to obtain. By the 1990s, only the Ministry of Defence had long-term research programmes with a given strategy, which allowed LETI to develop technology without the need to find research funding and industrial partners. However, more than 85 per cent of LETI's research has to reach an objective that has been defined in co-operation with industry directly or indirectly, which means that only 10–15 per cent of CEA funding is allocated for maintaining basic research. Most of the activity in LETI is focused on technologies three to seven years ahead of industry needs, but some work has immediate commercial value. For example, some of the prototypes, such as special circuits, were manufactured in line production. Licensing and patents are seen as the most effective way of getting a return on investment. In 1991 LETI held 400 patents, the majority were with larger companies, but some were with smaller ones working in niche markets for example in sensors. This reflects the second change which is that LETI has become more oriented towards SMEs.

France has much less activity in its science base directly related to flow measurement than the UK, although fluid dynamics is an important area of research in the university and national laboratory sector. The main centre of research relevant to the flow measurement industry is the Centre d'Etudes et de Recherches de Toulouse (CERT) which was established in 1968. It specializes in fluid dynamics and its research is

90 per cent funded by industry. It is a constituent organization of ONERA which was formed in 1946. Its mission is purely scientific and technical in the field of aerospace but has an additional industrial and commercial capability. It operates under the supervision of the Ministry of Defence, although only partially owned by the state.

3.6   Belgium: the university sector

Belgium has six full universities which have the traditional faculties such as applied science, law, medicine, philosophy, and research activities. These are the Université de l'Etat at Liège, Université Libre de Bruxelles, Université Catholique de Louvain, Rijksuniversiteit te Gent (RUG), Vrije Universiteit Brussel (VUB) and the Katholieke Universiteit te Leuven (KUL). The Université de l'Etat at Mons has some of the standard university faculties such as economic and social sciences. The engineering school at Mons is a separate institution. Ten other institutions of higher education have a limited range of courses. Approximately 80 per cent of students attend one of the six full universities; 54 per cent are in Dutch-speaking and 46 per cent are in French-speaking institutions. KUL is the largest university with 25,000 students which means that one in four graduates are from KUL.

Belgian state legislation sets out how universities are managed but the subjects taught are decided by the university. Ministries influence research by setting up large programmes such as biotechnology and the environment for which universities apply for funds. Universities lobby for subjects and ideas are taken from the universities and the government then decides on the programmes. One of the problems in recent years had been to find a political decision which will keep everyone happy. The solution found was to spread funding in proportion to their size across all universities. Centres of excellence have to have satellites in other universities (Van Geen personal communication 1991).

In the 1970s there was a big expansion of the number of students in universities in Belgium. The university population rose rapidly from 45,000 in 1966 to 100,000 in 1984. At the same time a decentralization policy was adopted with the objective of improving the take-up of university places. New universities were established at Kortrijkt and Limburg. The period of recession in the early 1980s saw the growth in university funding stop. At the time of federalization in 1980, Belgium had a national budget deficit and the national government did not transfer sufficient funds to pay for it all. This meant that the regions were obliged to build up a deficit. The total education budget (which funds the range of educational institutions from primary schools to universities) amounts to half the budget of Flanders. The deficit influenced policy: other social priorities such as national health, environment, and social welfare, were higher than university research. As a consequence, there were insufficient funds from the government for basic research, to equip laboratories and to attract top academics.

From 1980 onwards a series of transfer mechanisms were established to overcome the short-fall by increasing the flow of industry funding into the higher education system. Van Dierdonck *et al.* (1990) summarized the most important as:

   (i) industrial liaison officers at major universities;
  (ii) technological brokerage companies which can act as intermediaries between academic research and interested industrialists e.g. Leuven R&D;
 (iii) the technological guidance service of IRSIA (Institute for Scientific Research in Industry and Agriculture);
  (iv) research-orientated industry parks in close proximity to major university campuses;
   (v) technological innovation officers in the Walloon Region. These are scientists or engineers who are assigned to a company to carry out technological innovation projects. The EC and the Walloon Region financed 80 per cent of their salary. They are often closely linked to a particular university research laboratory;
  (vi) the creation of academic centres of excellence (Flemish Region);
 (vii) the organization of technology transfer days which bring together academic and industrial researchers in specialized fields.

As a result of the funding problems and the active technology transfer policy, in the last decade industry and the universities have become closer. Figure 3.1 shows the changing proportion of funding between fundamental and applied research at KUL between 1987 and 1991. It shows a dramatic climb in the growth of the 'applied research budget'; it is many times greater than the growth of 'fundamental research budgets'. The applied research budget comes from industry directly and through national and regional innovation support schemes. The pressure to find alternative sources of funding for research meant that academics accepted that they would have to undertake work for industry. The expansion in the system in the 1970s and 1980s meant that many new young professors were appointed. The combination of

MILLIONS Bf

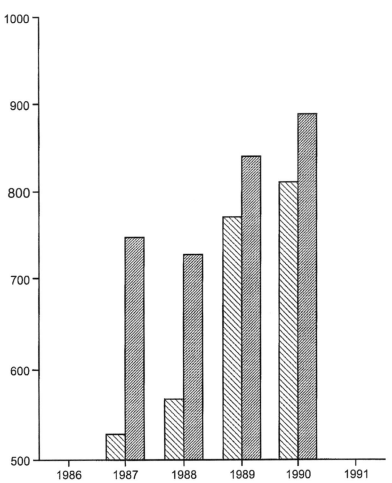

Applied Research = Research funded by industry directly and in conjunction with national and international agencies.

Fundamental Research = Funded by national science funding bodies.

Figure 3.1   Changing proportion of funding between fundamental and applied research at KUL between 1987 and 1991

the age profile and political realities produced a different set of cultural values towards research than those in the UK, which was better financed (Van Geen 1991 personal communication).

3.7 Belgium: national laboratories

Federalization devolved power to the regions which allowed greater freedom to develop regional strategies based on exploitation of technology. In Flanders technology policy in the 1980s stressed the development and application of advanced technology. IMEC was formed in 1984 by the Flanders government as part of its strategy of moving to a position of not being completely dependent on technology developed outside the region. The laboratory specializes in application specific integrated circuits (ASICs). IMEC was formed with a hundred staff and had quadrupled in size by 1994 (Lawton Smith 1995). It originated in KUL, which is located near Brussels. Partner universities are KUL, RUG and VUB.

IMEC's strategy is enacted along four major lines:

1 Scientific research in IMEC runs 5–10 years ahead of industrial needs, preparing technological possibilities which can be directed towards existing industrial companies. By collaborating with industrial top performers world wide, IMEC aims to build up its reputation as a 'centre of excellence' in the field of micro-electronics particularly in ASICs;
2 IMEC performs dedicated and flexible training in the field of VLSI chip design for educational as well as for industrial needs;
3 IMEC re-inforces the industrial research activities of Flanders in different ways:

   • by creating spin-off companies
   • by attracting foreign investment into the region by providing an investment package. To qualify for assistance companies must have R&D activity. This value added development activity is seen as encouraging anchorage to the region
   • by transferring technology into industry (for example in sub micron process technology)
   • by the supply of skilled personnel to local industry, as R&D staff leave at the end of their contracts.

4 IMEC carries out strategic research in collaboration with universities.

Universities, schools for industrial engineers and 13 high schools in Belgium are all linked to the central computer at IMEC. The Multi-Project Chip Service (MPC), supported by INVOMEC, is a low-cost application ASICs prototype fabrication service. Engineers work on test case designs and prototypes through the MPC service. From 1986 the MPC fabrication was extended to all universities in Europe, research institutes and industry. By 1992 over 500 designs had been fabricated. The majority have been for industry (214), followed by the Flemish universities (143) (IMEC Scientific Report 1992). INVOMEC is also active in the EC ESPRIT EUROCHIP programme providing training and software in design methodologies as well as training some 300 VLSI engineers per year.

SMEs can be given assistance with design and prototype develop-ment through consultancy and IMEC's servicing capability and test facilities. By 1991 IMEC staff had visited 110 Flemish SMEs to discuss their problems and how IMEC's expertise can be incorporated into the firms. Factors influencing the success of this strategy are cost effective-ness, ease of communication problems, psychological barriers, and mis-matches in the skill and technology bases of firms and that of IMEC.

Moves towards commercialization of the nuclear energy research centre, the SCK, include the formation of VITO in 1992. VITO split off from SCK in order to exploit developments arising from nuclear research. The laboratory specializes in industrial research on energy, the environment, biotechnology and advanced materials. It is a pub-licly owned organization, supported by the Flanders government.

5. Conclusions

The main features which differentiate the organization of the science base and the relationship with industry in each country are the means of the exploitation of national research assets. The UK public sector research base consists of a large number of institutions geographically dispersed through the country. The university system plays the dom-inant role in research but the national laboratories are also important but decreasingly so in particular areas of expertise. State funding for university research has declined steeply and there have been increasing levels of industrial and EU funding.

France has a different kind of PSSB from that of the UK. It has a nationally centralized but spatially concentrated pattern. This is focused particularly in the Paris region and in the South East of the country, for example in Grenoble. Its universities play a lesser research role while government funded laboratories, and joint university/ government research centres are the major players. State funding is more important than industrial funding of research.

Belgium contains elements of both the UK and French science bases. As in the UK, universities have a central position in the national system of innovation. To an even greater extent, there has been a long pattern of decline in state funding for research. The corollary of this is that industrial funding has exceeded state funding of research. On the other hand Belgium, like France, has moved towards centralized research facilities funded by the regions, particularly by Flanders, the most important of which is IMEC. IMEC like LETI is responsible for a range of interactions with industry including training and spin-out. Increasingly the Leuven/Brussels region is the focus of research and innovation in Belgium.

The main difference between France and Belgium on the one hand and the UK on the other, was perception within policy-making constituencies of the importance of developing clusters of linkages. Although there are local networks of innovators which link industry and universities but not the national laboratories, there has been no coherent strategy to identify the most appropriate means of using the resources of universities to encourage innovation in industry at the regional level. The case of IMEC is the most advanced example of this and provides a useful model of what is possible. In this case also the problem of competition between organizations within the same innovation system has been recognized and to some extent addressed by the major funding body, the Flanders government.

Part Two
Industrial Change

4
The Organization of Industry Filieres in the Flow Measurement and Electronic Component Industries

Introduction

The purpose of this chapter is to provide the background to the empirical evidence in Chapters 5 and 7 on the factors which influence the conditions under which firms in the two study sectors interact with universities and national laboratories. The first section discusses general factors which other studies have found to be important in firms' propensity to acquire externally developed technology. The second describes the main features of the two study sectors in each country. The final section draws these two strands together to summarize which factors would be likely to influence firms' participation in networking, collaboration and externalization activities with the PSSB.

1.   Sector and firm-level factors

Industrial firms both large and small are increasingly considering external technological acquisition in order to maximize their capabilities for further technological change. In *The Changing State of British Enterprise* (1996), Moore reported that there is a positive relationship between the growth performance of small- and medium-sized enterprises and their propensity to acquire new technologies externally. Moreover, acquisition of new technologies was more likely to be from other firms than from institutions such as universities. Other studies have identified a range of sector and firm specific factors which explain why some firms are more successful than others in acquiring externally generated knowledge. Sector-specific factors affecting the level and nature of demand include the structure of sectors including the relative importance of small and large firms; the degree of maturity of an industry; the representation of the interests of the sector in funding bodies; and the role that users play in the development of new technologies. Factors affecting the behaviour of individual firms include their ability to process and use information; changes to the internal organization of individual firms; size and patterns of ownership.

1.1   Sector factors

Intersectoral differences

It is on record that industrial sectors have geographical profiles intranationally and internationally which tend to persist over time. They are of greater or lesser significance in different national production and innovation systems and in the economic make-up of regions (Pavitt 1984; Archibugi and Michie 1995; Breschi and Malerba 1997). The structure, market orientation and R&D intensity of sectors may shape the propensity of firms to interact with external sources of technology.

Archibugi and Pianta (1992, 55–7) examined the technological specialization of advanced countries using patent data. They found that the UK does not have a clear pattern of specialization. Areas of greater strength include health-related fields, some chemical and mechanical engineering classes. It is generally weak in electrical areas, and more particularly in information and instruments. The most significant strengths of France are in state supported technological fields such as nuclear physics, medical preparations and weapons. It is one of the few countries that has a positive specialization in some electrical fields such as electricity, electronics and telecommunications. However, it is weak in other electronic-related technologies such as information and instruments and computing. Belgium's pattern of specialization is less uniform than in comparable countries, with generally weak correlations across different patenting institutions. Agriculture is the strongest sectoral specialization. There is a specialization in some chemical classes, such as inorganic chemistry and bio-chemistry. On the other hand, Belgium is strongly despecialized in the electrical and electronic classes.

Users' influence

Users not only determine the level of demand for outputs, they can also play a key role in the development of innovation filieres. Research by Von Hippel (1976), Dickson (1982) and David *et al.* (1994) on innovations in semiconductor and electronic process equipment, have shown that users often play a vital role in the development of new products. Other studies have found that customers are a major source of innovation in other high-tech sectors. A survey of 50 high-tech firms in Oxfordshire which included instrumentation, software and biotech firms, showed that the majority rated clients or customers as the most important source of innovation both inside and outside Oxfordshire. Links with customers within the UK (not including Oxfordshire) were of moderately to highly significant for 69 per cent of firms compared to 51 per cent for universities and national laboratories. The proportion was even higher for links outside the UK where 35 firms (75 per cent) said that they were significant (Lawton Smith 1998, 443), compared to 36 per cent with universities.

Interconnection of sectors is also a key aspect of the diffusion of product innovations from supplier sectors to user sectors (Malecki 1990, 156). Malecki makes the point that in older sectors, where improvement innovations are replaced with increasing frequency by 'pseudo-innovations' innovation ceases to be a leading influence on market position. The knock-on effect is that demand for externally generated technology, where it previously existed, is decreased.

Geographical distribution of innovators

As was suggested in Chapter 3, the geographical distribution of innovators may depend on the characteristics of the kinds of external knowledge which firms need to incorporate into their in-house innovative activities (Breschi and Malerba 1997, 142). They argue that proximity can be important where externalities are present resulting from innovative activities being generated from within a restricted group of specialists, for example in mechanical engineering districts in Italy or micro-electronics industries in Silicon Valley. Alternatively the spatial boundaries of knowledge might be dispersed if the knowledge relevant to the innovative activities is relatively simple, generic and non-systemic and the opportunities to innovate are embedded in generic and standardized capital goods, materials or intermediate inputs, as is the case of several traditional sectors. Thus the nature of the information can affect the kinds of external innovation relationships in which firms participate and the location of those activities.

1.2   Firm-level factors

Ability to acquire and process information

Innovation performance is associated with current levels of investment in R&D and the employment of staff of appropriate calibre to understand and interpret the value of academic research. Firms vary in their ability to recognize and incorporate breakthroughs in related technologies and respond to substitution technologies (Kleiknecht 1987, 123 quoted in Malecki, 1991, 168). Capacity is derived from a firm's level of prior knowledge and accumulated technological position (Dosi 1984, 15). Also a firm's position around an industry average technology trajectory may be important in its external relations (Bye and Chanaron 1995, 63). Its capacity to interact is dependent on appropriate management strategies which include commitment to, and visible support for, innovation on the part of top management, long-term corporate strategies in which innovation plays a key role, and a long-term commitment to major projects (see Rothwell 1992, 227). Key individuals who are able to process information from external sources have been identified as 'technological gatekeepers' (Allen 1977). In order to fulfil this role they participate in personal networks and technological communities to which entry is based on education and professional experience (see Angell *et al.* 1985; De Bresson and Amesse 1991; Cohen and Levinthal 1990; Debackere and Rappa 1994).

The experience of interaction is also a factor in whether firms engage in externalization of innovation. Research has shown that interpretations of successful behaviour in the past will continually be reproduced and reinforced as long as they seem reasonably efficacious (Maskell and Malmberg 1995, 5). Failure can discourage interaction even though there may be potential gains.

Firm size

The innovativeness of small and large firms appears not to be constant between sectors or countries. The 1994 OECD Report on Science and Technology Indicators showed that SMEs rank among the most and the least technologically progressive of companies. In the large firm sector, Europe is distinguished by the fact that its very largest companies are smaller than the comparable US and Japanese firms in terms of both research and sales. However, among those that are simply large, European enterprises fare better in sales and R&D levels, often achieving a higher level of research intensity than comparably sized firms in the US and Japan.

The Report showed that research into factors influencing small firms' participation in innovation networks has found some persistent country-specific patterns. In the UK, SMEs' share of R&D employment exceeds their share of R&D expenditure, which suggests that R&D activities are more labour intensive than in large enterprises. Belgium also has a relatively high share of SME R&D for its level of expenditure. Moreover, the absolute numbers of firms in each size category is as important as the relative proportions of small to large firms. For example, France and the UK have the smallest SME sectors across the EU but the influence of innovative small firms may be felt more strongly in sectors where they are more numerous than in sectors where there are fewer.

Ownership

The location of decision makers who decide where R&D will be undertaken in multi-plant firms is of major significance in the construction of innovation filieres. R&D intensity of indigenous firms compared to foreign-owned firms is an important variable. In some countries, (for example UK and Belgium) foreign-owned companies account for a high proportion of innovative outputs (Patel and Pavitt 1990 in Walker 1993, 174/175). These authors argue that a growing part of the British innovation system has become an appendage of foreign-owned systems. This comment equally applies to Belgium but not to France (see Chapter 2). Moreover, as indicated by patent data examined by Patel and Pavitt, British firms are comparatively highly internationalized in their R&D activities.

Representation

Firms' propensity to interact with the PSSB is also affected by the possibilities of acquiring technological input from other industrial sources such as specialist trade research organizations such as the UK's SIRA and PERA and trade associations. Trade associations are the formal representation of an industry's identity and are usually, although not always, nationally constructed. Their importance as sources of technology varies by country. In the US for example, they have been shown to be important in organizing innovation. Moreover, although they have little direct power over companies, they 'allow an industry to present a united political front, undertake joint research projects, provide information and services to their members, and engage in industry-wide labour recruitment, wage bargaining and strike management' (Storper and Walker 1989, 136).

2. The flow measurement and electronic components industries: definitions and industrial structures

This section begins by defining the two industries. Then follow details of markets; a description of the sectors in each country; the geographical distributions of firms within each country; and lastly drivers of technological change. The section shows that both sectors are dominated by MNCs but independent, often small, firms play important but different roles as innovators. In both sectors, the UK is the dominant producer.

The flow measurement industry

The flow measurement industry is technically diverse. There are over 100 flow meter designs. These have been categorized into 12 classes in the British Standard on the Selection and Application of Flowmeters, based on principle of operation. Positive displacement (PD) and differential pressure (DP) meters account for nearly 40 per cent of the UK market, but DP meters have the largest market share in Western Europe. Despite the large range of meters available, process flow measurement is dominated by differential pressure producers using technologies such as orifice plates and venturis with associated differential pressure transmitters. It is estimated that at least 60 per cent of all flow measurements are undertaken using these techniques.

In the UK the industry is classified under 1980 SIC categories of 3710 'Measuring, checking and precision instruments and apparatus', and 3442 'Electrical instruments and control systems'. Meter production is divided into two parts: industrial process activities and domestic utilities – predominantly water and gas. In the domestic field, metering of gas, and in most countries, water is a standard billing procedure. Many firms make a range of level and pressure instruments and flow measurement is only part of their business activity.

The UK is the largest location of both manufacturers and suppliers of flow meters. In the early 1990s there were some 200 flow measuring companies in the UK. Some 80–100 were manufacturing companies and the remaining 100 were agents (Furness and Heritage 1989). Other commercial sources indicated that France had at most 30 and Belgium less than 20 firms, mostly sales offices of multinational companies. Employment in the UK flow measurement industry at 81,500 exceeded that of electronic components in 1991. The figures should be treated with some caution. The figure for electronic components is likely to be more accurate than for flow measurement as the SIC codes include activities other than flow measurement. The largest companies are Rosemount (US) and ABB-Kent (Sweden). Acquisitions include Schlumberger's (US) purchase of EMI (UK). In a reverse pattern, 1991 the UK company Siebe acquired the US industrial controls group Foxboro. Remaining UK firms include United Gas Industries (UGI owned by Hanson), and Mowbrey.

Flow measurement is a highly competitive market with many suppliers and low profit margins. In the late 1980s the world-wide market for meters was in the order of £450–500m per annum. Among the eight major Western European countries which account for some 85 per cent of the market, in 1991 the UK was the second largest user of flow instrumentation with a market share of 19.0 per cent, (behind Germany 30.9 per cent), France was third (16.5 per cent), and Belgium seventh (5 per cent). Belgium was the second smallest user of instrumentation in Europe. Commercial sources indicate that consumption of flow, pressure and level instrumentation in Belgium was $43.8m in 1986 and was expected to rise at an average of 6.4 per cent per annum to reach $64.1m by 1992.

The UK has a small percentage of the world market in industrial flow meters which is somewhere between £0.5 and £1 billion a year. On the other hand it has the largest domestic market in Europe, which is equal to the industrial market. The total turnover of UK firms in 1988 was estimated at £90m with 96,000 devices sold. In the domestic meter market, the UK market for gas meters was two million units per year, compared to 400,000 in France and 60–80,000 in Belgium. In France, gas is a much smaller source of energy than nuclear electricity. Water meters have had a constant share of the UK market with around 300,000 each of domestic and industrial water meters sold in the UK, worth about £12m a year, with an equal share of the market. The value of manufacture of meters in 1993 was £130m of which some £56m was exports. In Belgium, most energy is nuclear electricity (60 per cent), hence there is a small market for gas meters. The low usage of instrumentation results from a combination of a small domestic market, the underdeveloped national measurement system and the low priority of accurate metering in some markets (see section 4).

The value of flows measured can be up to 6 per cent of GDP of an advanced industrial economy (Quinn 1994). Chemical and petrochemical industries constitute the largest market, which is twice as large as the next category, power generation. Other major users are petroleum refining, water and effluent and primary metals industries. Smaller markets include food and beverage, and pulp and paper. The UK value of oil and gas monitored in the mid-1980s was in excess of £20 billion per annum.

In this industry, the top six companies share just over half the world market. The process of restructuring in which the dominance of larger firms has increased through hostile take-over of smaller independent firms has accelerated since the mid-1980s. Most of the remaining innovative smaller firms are independent, family-owned businesses based on mainland Europe, which are not vulnerable to acquisition.

In France and the UK, two manufacturers dominate the domestic meter market. Schlumberger is one of the two in the UK and France and is also by far the dominant manufacturing company in Belgium supplying 80–85 per cent of the gas meter market. The Belgian industry is largely dominated by the sales offices of multi–national companies. Some 60 per cent of the country's market is taken by four companies (three European and one US); the remaining meters are imported.

2.2   The electronic components industry

The industry is classified into ten major product groups: capacitors, connectors, discrete semiconductors, integrated circuits, printed circuit boards, resistors, switches and relays, and wound component and materials (NEDO 1990), the largest of which is integrated circuits. The UK electronic components industry falls within UK 1980 SIC categories of 3444 'Components other than active components, mainly for electronic equipment', and SIC 3453.1 'Active components'. The latter includes memories, microprocessors, micro-controllers, ASICs and the like. In France the industry is defined by NAP sectors 2915 (passive components) and 2916 (semiconductors) (Swyngedouw 1992).

Data on the number of firms and employees in the electronic components industry was collected from official, academic and trade association sources. They show that the UK has the most firms and more employees than France, and both have considerably more of each than Belgium. France has a higher representation of larger firms in the industry than the UK (Breschi and Malerba 1997, 153).

In the late 1980s the UK had some 700 manufacturing sites, employing 90,000 (DTI 1990b). Moulaert and Swyngedouw (1992) found that France had over 600 manufacturers and suppliers, employing some 40,000. Figures supplied by the Belgian Trade Association Fabrimetal (1991) showed that Belgium had about 25 manufacturers and suppliers. Employment in the UK electronic component industry had been steadily decreasing during the 1980s, down by 1.6 per cent from 103,000 in 1980. NOMIS data showed a much smaller total employment for 1991, just under 60,000. This has in part resulted from some consolidation of activity. For example GEC Plessey Semiconductors (GPS) was formed by the merger of Plessey Semiconductors and Marconi Electronic Devices Lincoln (MEDL) in 1990. In contrast, in France, Moulaert and Swyngedouw found that employment was increasing.

Table 4.1 EEC electronic components
markets in 1990

Country	Amount ($)
W. Germany	7154
UK	5475
France	4676
Italy	2928
Netherlands	1291
Spain	1270
Irish Republic	915
Belgium	734
Denmark	218

Source: DTI 1990a

In the European electronic components industry in 1990 Germany was the largest European market, the UK second, France third and Belgium eighth (Table 4.1).

In 1989, the UK produced electronic components accounted for 7 per cent of inputs compared to 22 per cent of imported electronic components (NEDC, 1991 7). Over 50 per cent of output was exported whilst nearly 70 per cent of component requirements were imported. Overall there was a large and growing deficit amounting to some £1.5bn in 1989 (NEDC 1991, 2). This is associated with the UK's specialization in electronic end-products, especially military and industrial applications (NEDC 1991, 6).

France had a negative balance on exports in electronic components throughout the 1980s. In 1987, the ratio of exports to imports was 0.87 (Chesnais 1993, 225). Belgium is a fast-growing market with imports and exports accelerating. It has a positive trade balance on electronic components. In overall trade Belgium is the UK's third largest EU market when Luxembourg is included (DTI 1990a).

Countries vary in the extent to which parts of the industry is represented. At the 'leading edge' market in electronic components, semiconductors, EU countries have a limited presence. Global production of semiconductors is highly concentrated with the top ten firms producing over 50 per cent of output. Europe's competitive performance in semiconductors has been poor by any standards (Hobday 1991, 81). In 1987, ten Japanese (including NEC, Toshiba, Hitachi, Fujitsui, Mitsubishi), seven US and three European firms accounted for 70 per cent of world-wide semiconductor production. The top European firm is the Dutch company Philips-Signetics (11th) followed by SGS-Thomson (Italy/France). INMOS and Plessey in the UK were then important in niche markets (OECD 1992, 137). European firms had still not achieved the critical threshold of 5 per cent of the world market. Japanese firms account for nearly 90 per cent of world production of high capacity memories, while US microprocessor manufacturers (Intel, Motorola) control over 80 per cent of world production of 16 and 32 bit microprocessors. Japanese firms now dominate DRAM production, having greatly 'increased market share through investment in equipment and research, an emphasis on mass production techniques, exchange rate advantages and the ability to produce at lower costs' (OECD 1992, 133). In the electronics industry as a whole, all European countries faced deteriorating trade balances between 1978 and 1986, but the greatest decline was experienced by Britain (Walker 1993, 168/9). At the other end of the market, the UK demand for PCBs in 1991 was worth £372 million while total consumption was worth between £430 and £465m. The UK market represents about 29 per cent of the European and 7 per cent of the world market for PCBs (Turok 1993b, 1791).

The system of production in the industry varies between the three countries.

The relationship between producers and users

The UK and Belgium have no domestic consumer electronics industries. Thorn, the last UK television producer, was acquired by Thomson of France. As a result, there no longer exists a pattern of vertical integration between user and producer. This has meant a reduction in the need for the major electronics companies to maintain stand-alone semiconductor production. GEC, Plessey and STC, as well as INMOS and Philips at Redhill, have ceased the production of silicon chips.

Location of control within the industry

While France has a high proportion of French owned firms, some partly state-owned, the UK has high levels of inward investment and has generally failed to protect its own interests in components. For example INMOS has been sold to SGS-Thomson (Walker 1994). The INMOS name disappeared in October 1994. The UK has the highest levels of foreign-owned semiconductor firms in Europe, amounting to some 30 companies, more than the rest of Europe put together. They are mostly Japanese. Between 1984 and 1992 Wales attracted 16,833 manufacturing jobs in electronic components, more than 10 per cent of the UK total, compared to 35,209 in Scotland, and 35,281 in the South East (Huggins 1995, 11). The attractiveness of the UK has been explained in terms of:

* absence of hostility to overseas firms
* a good infrastructure
* support from the DTI and local authorities
* the English language, important for Japanese firms
* low wage rates
* good industrial relations (NEDC 1991, 3/4).

Scotland is dominated by foreign firms which accounted for 58 per cent of all electronics jobs (Turok 1993b, 1789/1792). Semiconductor manufacture is a major section of the industry in Scotland (Turok 1993b). A small sub-sector is printed circuit boards (PCBs) which account for around 4 per cent of the total in electronics. Scotland has four specialist PCB-assembly plants: Avex, Phillips, SCI and Timex. They have highly sophisticated PIH machinery and up-to-date SMT equipment. All are foreign-owned and employ about 2500 altogether (Turok 1993a, 407).

Type of R&D

The third difference, which arises from the second, is the kind of R&D undertaken by domestic firms. For example, Walker (1993, 175/6), quoting Cabinet Office statistics, shows that over the period 1975 and 1986 there was a very marked growth in R&D in electronics, and an apparent higher proportion of R&D allocations to electronics in Britain than in other cited countries (for example France and Germany). However, Walker finds this something of a puzzle as individual firms such as GEC and Plessey had not been an area of competitive advantage for Britain. He cites evidence which shows that Britain's technological advantage in electronics slipped over the period when R&D expenditure was increasing. His explanation for this lies in the extent to which R&D in electronics in 1985 was conducted by multinational companies (about a half of all multinational spending on R&D) and on R&D for defence purposes (about a third of all electronics R&D). Therefore just over half of electronics R&D may have been carried out by British firms orientating themselves to civil markets. Moreover, this civil expenditure seems to have yielded a low return in terms of exports and economic output. The largest civil item, the telecommunications switch, System X, was not exported and British firms have been generally unsuccessful in high-volume areas such as semiconductors and consumer electronics.

There is then a declining share of electronics produced in the UK which have been designed in the UK. The proportion was expected to decline from nearly 60 per cent in 1988 to about 50 per cent in 1995. Productivity in foreign-owned firms was 20 per cent higher than domestically owned firms. Growth in output and productivity in the components sector of electronics (at 6.2 per cent a year for the period 1980–8) was below that in the sector as a whole but above manufacturing as a whole (5.3 per cent). However, at the niche market research end where investment costs are lower, for example in gallium arsenide (GaAs), formerly indigenous firms are first rank. UK electronics firms in Scotland – such as Ferranti, Marconi and Racal – have been preoccupied with small batch, highly priced products for protected military markets (NEDC 1991). In France the sector also has been the target of foreign capital. Storper and Salais (1997, 131) find that foreign presence is high in semiconductors. Import penetration is around 35 per cent in components. Almost 70 per cent of all electronic imports are components or core technologies. The weakness in French electronics is in components but the strength is in large-scale systems which are frequently ordered by civilian or military bureaucracies.

Belgium has also retained a share of small indigenous firms, but the larger firms are in foreign hands, for example Alcatel, which acquired MIETEC in 1993. The pattern of ownership in Belgium is shown in Table 4.2.

Table 4.2   The Belgian electronic components industry showing number of firms by ownership

Country	No. of firms
Belgium	13
UK	1
Germany	2
US	3
Dutch	3
French	2
Japanese	1
Total	25

Source: Fabrimetal

3. Geographical distribution of the industries

The main similarity between the industries in all three countries is that employment is concentrated in core regions. However, there is more evidence of clustering of innovative activity in France and in Belgium than in the UK. Breschi and Malerba (1997, 153) found France showed a higher level of geographical concentration of innovation activities than the UK in the electronic components sectoral innovation system and in mechanical and electrical technologies.

In the UK, the South East contains a quarter of employment in flow measurement and over a quarter of electronic components employment (Tables 4.2 and 4.3 and Figures 4.1 and 4.2). Scotland had a higher percentage of electronic components employment (17 per cent) than flow measurement (8 per cent). Wales had 10 per cent of employment in electronics but only 1 per cent in flow measurement.

In France there is some centralization of activity in Paris and in the South and South East in both sectors. Production in electronic components is centred on four main areas – Paris, Centre, Pays-de-la-Loire, and Rhône-Alpes as shown in Figure 4.3. The industry experienced a large net growth in employment which was concentrated partly in Paris, but with additional growth in the south and west, where labour

Table 4.3 Employment in the flow measuring industry and the electronic components industry September 1991 based on SIC (1980) activity headings

	Flow measurement headings 3442, 3710	Electronic components headings 3444, 3453
Region	*Persons*	*Persons*
South East	21 100	17 100
East Anglia	5 800	2 400
London	6 600	3 400
South West	9 500	7 000
West Midlands	8 100	2 000
East Midlands	6 300	2 900
Yorkshire and Humberside	3 000	1 500
North West	8 100	3 400
Northern	3 900	4 300
Wales	2 200	5 900
Scotland	6 900	9 900
Total	81 500	59 700

Source: NOMIS Government Statistical Service
NB The totals do not add to sum of column figures due to rounding

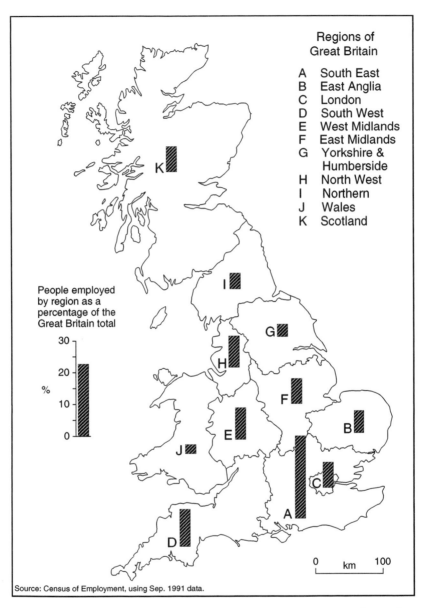

Regions of
Great Britain

A   South East
B   East Anglia
C   London
D   South West
E   West Midlands
F   East Midlands
G   Yorkshire &
    Humberside
H   North West
I   Northern
J   Wales
K   Scotland

People employed
by region as a
percentage of the
Great Britain total

30

20

%

10

0

0   km   100

Source: Census of Employment, using Sep. 1991 data.

Figure 4.1   Employment in the UK flow measuring industry as at September 1991 based on SIC (1980) activity headings 3442, 3710

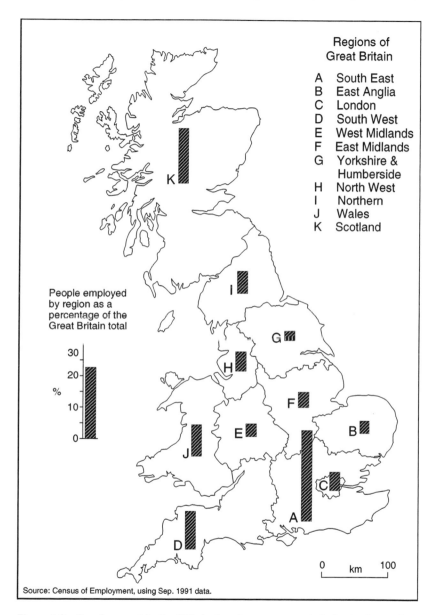

Figure 4.2 Employment in the UK electronics components industry September 1991 based on 1980 (SIC) activity headings 3444, 3453

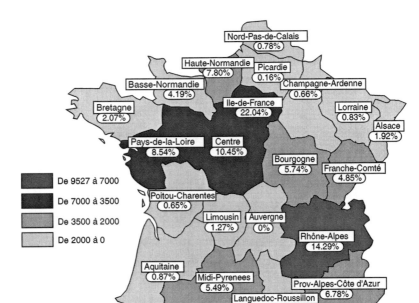

Figure 4.3   Location of production in the French electronic components industry, 1987, NAP sectors 2915 (passive components), 2916 (semiconductors).

Source: Swyngedouw, Lenattre and Wells, 1992.

costs are lower. The dominance of Paris is much less pronounced and
the peripheral regions are doing reasonably well (Swyngedouw *et al.*
1992, 110, 114). On the other hand, for certain technologically
advanced sections of the electronic components industry, the Ile-de-
France is almost the only location in France; for example 34 of the 43
firms designing ASIC circuits are located there and only nine outside it
(Storper and Salais 1997, 138).

   In Belgium, Flanders contained the majority of R&D activities of the
electronic components industry, including some specialization in telecoms, while most of the production sites were located in Wallonia. In
flow measurement, only three manufacturing companies were found,
two in Wallonia and one in Flanders and there was a small number of
sales offices in both Wallonia and Flanders.

4. Technological change

Sectors are rarely uniform in the rate of technological advance across the range of products which comprise the industry. The difference between the two sectors in this study is the relative conformity in the rate of technological change within the flow measurement industry and the relative diversity in electronic components. At the same time there has been a tendency in both sectors for small firms to move into and develop specialist niche markets on the basis of their innovative approaches.

4.1 Flow measurement industry

The flow measurement industry is characterized by a slow pace of technological change. It is a 'low-alpha' industry – one in which the returns from out-spending rivals to improve the technical performance of any particular product are limited (Sutton, forthcoming). Three factors inhibit the propensity of firms to innovate and hence the likelihood of them collaborating with the PSSB.

A. The long life expectancy of meters and the high costs of flow-meter development

The expected life of a meter design is ten years. The maturity cycle for a new flow meter can extend to up to 20 years. For example, electromagnetic flow meters which first appeared in the late 1950s are now widely used and accepted but this acceptance has only come in the late 1970s with associated developments in field techniques (Sanderson 1989, 1). Their simple construction makes them suitable for use as 'hygienic' meters for use with food and drink (Hayward 1991, 29). In the 1980s vortex meters became popular. They operate on the principle of disturbance. Something is fitted inside the pipe to create a periodic disturbance within the fluid, and the frequency of the induced disturbance is taken as a measure of flow rate.

The development of the coriolis mass flow meter (vibrating tube) was taken up in the early 1980s, and established by the 1990s. In 1983/4 there were only two manufacturers of these meters; by 1990 there were some 14. The meter is being developed making use of electronics and a very old principle of coriolis force. Its advantage is that it answers the significant problem of measuring mass-flow rate as opposed to volumetric-flow rate. It is in use for example on industrial flows of fuel oils and there are proposals to introduce them to petrol pumps (Johnston 1993, 10). Like all meters, it has its limitations. The present meters are not suitable for situations subject to even moderate levels of vibration. The first device was the *Micromotion* meter, still the most widely used, in which the flow is made to pass around a pair of cantilevered U-tubes. The majority are used in process control purposes, mainly with liquids, giving extremely accurate results. Applications include off-shore oil and gas industries and the food industry.

The reason for the delay is that it takes time to build up confidence in a new technology and for testing to be completed. Developments in the field are welcomed with high – perhaps too high – expectations, with the resultant disappointments, delays, extra maintenance effort etc. (van Kampen 1989). Under pressure of competition, manufacturers are increasing compelled to introduce novel or improved instrumentation, 'it is a known fact that salesmen attempt to sell products straight off the drawing board to steal a march on their competitors...' (van Kampen 1989, 2) which means that factory testing is not always undertaken thoroughly. The costs to the user of equipment which does not meet expectations include delays in the construction of new plant, adverse effect on product quality, and less than optimal operation of the production process.

By 1991 it was possible to argue that, owing to the wide variety of flow meters available, there were then few problems left in flow measurement. Exceptions are that of accurately measuring multiphase flow, and into the installation of meters in non-ideal conditions. However, even the former was expected to be solved by research being undertaken at that time (Hayward 1991, 31). An example of the latter type of problem is the design of water meters. In 1993, the results of the National Water Metering Trials were published. Trials were undertaken in 11 areas. On average 20 per cent of meters removed at random from test sites and tested failed to meet the in-service test requirements.

Experts in the field of metrology consulted during the study considered that there is little prospect of a new concept emerging in flow metrology. Development of existing or established concepts will be driven by market requirements. For example, there was thought to be a huge potential for flow meters which can be fitted into engine management systems for motor cars. In addition, in the North Sea Oil and Gas industry, with the increase in the use of unmanned production platforms and the move into deeper waters, the development and deployment of multiphase metering systems has become a major focus. The potential cost saving was well recognized although the technology was still in its infancy. The most important technological developments in the industry were likely to come not from within the industry but from developments in electronic components. The microprocessor revolution is radically changing the design of process instrumentation; developments have reduced the market life of many instruments from more than ten to less than five years. The meter itself may become redundant as the pipe through which gas and liquids flow becomes an 'intelligent' instrument.

An example of the importance of electronic technology in radical change of meter design is the gas meter. Advances in technology, particularly in microelectronics, spurred British Gas to reappraise the metering options for the domestic market. Technical specifications for a new meter were announced in March 1987, and companies were invited to submit their ideas. In September 1990 British Gas announced that it had chosen two ultrasonic meter design companies. These were Gill Electronics, a small independent electronic research and design company employing 25–30 people, and FML, a subsidiary of Siemens, based at Oldham (Cookson 1990). British Gas owned the IPR to the Gill Electronics meter. Neither of the existing two main suppliers were chosen for the programme. An electronic meter designed by Schlumberger was one of five short-listed in 1988. It was withdrawn after some initial development work when its development team failed to meet the deadline for the next stage of evaluation. The company subsequently licensed the Gill technology. UGI had approached a university with a view to working on a new meter design. Gaz De France had gone through a similar process in designing a new meter for British Gas and had selected a small number of firms, including Siemens; by 1992, this had not been completed. The development of the new UK gas meter illustrates a gap between users and the universities, users and manufacturers, and ultimately the limitations of FLOMIC.

B. The recession of the late 1970s and 1980s which reduced the demand for meters

The recession left process industries with over-capacity causing intense competition between producers within major markets inside and outside Europe, mainly between Europe and the US and Japan. This had an important impact on the structure of the market. Larger companies generally were selling complete systems, which included software as well as instrumentation. They tended to invest in the systems aspect of measurement rather than instrumentation. Instrumentation was sold on the coat-tails of the turn key systems. In the depths of the recession, older plants in process and petro-chemical industries were refurbished and upgraded as firms adopted a strategy of cost cutting rather than investing in new plants. Smaller, particularly independent mainland European companies (for example from Switzerland and Germany), identified the weakness of the instrumentation component of measurement systems and began to develop significantly better instruments. This caused fragmentation and specialization in the instrumentation part of the market as niche markets developed. This meant that when users began to look for better performance from instruments they bought from the smaller rather than the larger firms. This in turn opened the UK market to a potential threat from more innovative Japanese and from German companies. At the same time competition on price rather than performance became more intensive as users also demanded cheaper and sometimes simplified instruments.

C. *Withdrawal of users from meter design*

Technological change in the flow measurement industry until the late 1980s went hand in hand with the involvement of users in the design. However, users of flow measurement sensors have tended to be very conservative in the choice of metering method (Sanderson 1989, 1). This conservatism has been exacerbated by another trend, that of the withdrawal by industrial users from meter design. This meant that the costs of innovation were then borne entirely by the manufacturers.

In the late 1980s and early 1990s there was an increasing separation of activities. In the UK and France, this relationship has been redefined. The general trend is clear. Until the early 1990s, user companies, such as BP, used to maintain laboratories and teams of researchers in the particular field of flow measurement which they saw as critical to the operation of their plants. Since then many have chosen to rely on manufacturers and conformity to standards to provide the flow measurements they require. The same economies have also taken place at individual process plants with a resulting reduction in depth of flow measurement technology on site. This, taken with a similar reduction in the depth of assistance available from manufacturers, may prove to be one of the limiting factors in the rate at which some of the new technology can be introduced (see Mason and Wagner 1994).

The International Instrument Users Association, the WIB, formed in 1963, has members in nearly all European countries, the United States and Japan. Its aim is to 'arrange, at the joint expense of members, for independent testing of the performance, quality and reliability of instrumentation and control systems that can be used in industrial processes'. Similar organizations are active in the UK–SIRA

Instrumentation Panel (SIREP) – and in France – Association des Exploitants d'Equipements de Mesure, de Regulation et d'Automatisme (EXERA), and close ties exist between these organizations. The Summary of Evaluation Findings on flow, level and pressure instruments evaluated by WIB, SIREP and EXERA during 1989–93 showed a very low level of accuracy, only 22 per cent of instruments met all specifications. Over three-quarters failed manufacturers' or users' specifications, of which over half failed manufacturers' specifications. The major problem facing manufacturers was the influence of external conditions on meters post-installation.

4.2 The electronic components industry

The electronic component industry covers a wider range of technological sophistication and of research intensity than the flow measurement industry. At one end of the spectrum are low-tech producers such as PCB manufacturers which generally undertake little R&D themselves (Turok 1993b, 1796) and at the other are high-tech SMEs, for example those which make materials for the semiconductor industry. In the 1980s, R&D in UK electronics had a very high ratio of value added, exceeded only by the US. In 1986 the figure for electronic components was 12.2, compared to 15.2 for the USA, 6.0 for Japan and 10.4 for Italy (NEDC 1991, 8). The NEDC reported that the lead held by US firms in the early 1980s had been surrendered to the Japanese. European and other firms remained largely dependent on technology licensed from American companies.

Four dominant technological trends in the electronic components industry in Europe in the late 1980s and early 1990s were identified by the DTI (1990b).

1 The defence sector in Europe would decrease by an estimated 8 per cent between 1989 and 1995.
2 There would be an increase in the consumer industry and telecommunications, and a dramatic rise in automotive markets.
3 Users would continue to demand smaller, faster products with greater computing power.
4 Applications of both silicon and gallium arsenide technology would continue to increase.

The dominance of integrated circuits as the building blocks of modern electronic systems has led to an intense international competition to develop standardized products (such as DRAMs and microprocessors) that can amortize the high cost of R&D and the dramatic increase in capital intensity of state-of-the-art production facilities. At the same time, electronic system producers have increasingly turned to specialized ASICs to create technologically differentiated and innovative products (OECD 1994, 84). Custom devices, especially ASICs, which accounted for 15 per cent of semiconductor sales in 1988, were the fastest growing type of integrated circuit. ASICs were forecast to amount to 25 per cent of world semiconductor production by the mid-1990s. These specialized, design-intensive ASICs markets offered opportunities for small European firms to prosper.

Users have also been important drivers of technological developments in the electronic components industry (Turok 1993a, 402). Here, the impact has been more positive than in the flow measurement industry. The automotive, data processing and telecommunications sectors now place much more emphasis on innovation, research and development than other sectors as these are sectors firmly established, technological change is rapid and joint product development is seen as a key success factor. On the other hand, as mentioned earlier, the structure of the market has had a limiting effect on the application of R&D in civil markets. In telecoms and defence electronics, monopsonist purchasing arrangements in the past have been dominant factors in shaping the direction of R&D. In the UK, the cost plus basis of sales, plus the retention of IPR by MoD has meant that companies have limited the need for strategic thinking outside these two markets (DTI 1990b, 25).

5.   Regulations and regulatory change

The differences in the national and international regulatory environment are important influences on sectoral patterns of innovation. Regulations include national and, in Europe, EU technical standards and environmental regulations. Wider regulatory change in the UK includes the privatization of utility companies and firms in the public sector.

5.1   Standards

Although compliance with standards is a major production criterion in any industry, it is a high technical priority in the flow measurement industry. A survey carried out for the UK's DTI (Brookes 1994b) indicated a growing demand for accreditation to ISO 9000, the international standard for quality. In order to achieve this, traceable measurement must be demonstrated which means that firms are obliged to devote their engineering efforts to meet the regulations. The industry is regulated by seven EC regulations regarding measurement for fiscal purposes and on standards, such as on environmental quality. The CEM (an EC body) is working to reach international agreement on standards.

National differences between the methods by which standards are set and regulated also play a part in the innovation and competition strategies of firms. In flow measurement, the UK has a far more advanced standards infrastructure than France and Belgium. In the UK, the DTI administers the National Measurement System (NMS) through the National Measurement System Policy Unit. The existence of NMS enables the maintenance and dissemination of measurement standards and the continuance of research into new and improved standards. The National Measurement Accreditation Service (NAMAS) accredits, against national standards, the competence of laboratories, for example those owned by manufacturers, to carry out calibrations and other measurements. The National Engineering Laboratory (NEL) has the responsibility to the National Measurement system of maintaining and disseminating UK fluid flow standards. All gas and oil fields have their own metering system and companies are obliged to report production figures to the oil taxation office, part of Inland Revenue. Companies are taxed on those figures, and therefore accuracy of meters is of financial importance.

In France, scientific metrology is the domain of the Bureau National de la Metrologie (BNM), which receives funding from the Ministry of Industry. Industry is represented on BNM's management committee, which is composed of representatives from various ministries and the primary laboratories as well as industry (Brookes 1994b, 37).

In Belgium, the policy unit, the national measurement system as controlling body and the national standards laboratory are all functions of the Inspectorate General of Metrology (IGM), which is part of the Ministry of Economic Affairs. It has responsibility for calibration. It does not conduct research and there are no specific metrology research programmes in Belgium. Moreover, there is little or no dialogue between the IGM and industry in Belgium and no advisory committee exists (Brookes 1994b, pp. 66–8).

5.2 Privatization of utility and state owned companies in the UK

Privatization of the utilities in the UK changed the structure of R&D in both industries. This an area of research where more analysis of the consequences of privatization is needed, especially in the electronic components industry, for example of the impact of privatization of BT in 1984 and BAe in 1981. Rather more information is available about the flow measurement industry. Pre-privatization, the utility industries exhibited classic vertically integrated structures with regional research centres. Since privatization, utilities have been subject to intensified performance targets, due to changing regulatory and competitive environments. New pricing policies set by the regulators influence profitability, and therefore potentially, R&D spend. Major upheavals in the structure of client industries have occurred due to a series of mergers and acquisitions. These have been both by firms in the same industry (for example Scottish Power's acquisition of Manweb in 1995), and across industries (for example North West Water's acquisition of Norweb the same year).

An example is the impact of privatization on innovation strategies in flow measurement in the water industry. Between 1974 and 1989, R&D in water companies was mainly undertaken through a subscription scheme, based on a percentage of turnover, with the work being done by the Water Research Centre (WRC) at Medenham. Since privatization, water companies have been able to determine their own R&D direction. The effect has been that there is much more secrecy within the industry and less sharing of technological information. However, annual accounts of the water companies show an increasing R&D spend. Privatization of the water industry in the UK in 1989 released some £28m for investment by the water companies, and removed operating constraints on investing overseas. Some companies have diversified into instrumentation. This could shape future technological developments through the power of these companies being both major users of and investors in new meter design.

6.   Representation and industrial coherence

A further country- and sector-specific factor in the focus of innovation activities is that of trade associations. These organizations were included in the study because they have been found to be players in the innovation process in the US. In the three European countries studied, they vary in the contribution they make to the organization of innovation.

6.1   The flow measuring industry

The trade associations and organizations representing the interests of flow meter manufacturers are GAMBICA in the UK, UDIAS and

Fabrimental in Belgium, Symecora in France and at the European level, EUROMET. GAMBICA is active on its members' behalf in ways related to innovation: for example it informs members about government schemes for interaction such as those under LINK. The UK has other organizations set up to support innovation in industry, notably the commercial organization, SIRA. UDIAS is an organization of sellers of instrumentation. It runs exhibitions and occasional study days.

Flow Measurement Instrumentation Consortium (FLOMIC)

The UK-based organization FLOMIC was launched in 1986 and established on 1 January 1987. It was based at the department of Fluid Engineering and Instrumentation at Cranfield. The objective of FLOMIC was to fund research at the front end of flow measurement, and to act as part of the intelligence network, linking academic and industrial research. Its formation was spurred by earlier efforts in the 1980s to encourage collaborative research programmes between industry and university departments which had failed. It held a register of expertise in universities, updated every two years. FLOMIC was given pump priming funds initially by the DTI but until its demise in May 1993, was funded by subscription. At its peak it had some 25 industrial members, users and manufacturers plus ten academic members and the National Engineering Laboratory (NEL). Its membership was international through corporate membership. The academic departments represented were:

- Teeside Polytechnic
- University of Bradford
- University of Liverpool
- University of Manchester
- University of Salford
- University of Sheffield
- University of Southampton
- University of Surrey
- University of Sussex
- Cranfield Institute of Technology.

Initially FLOMIC recruited large firms from major industries, water, oil, gas, iron and steel. It was less successful at recruiting medium- and small-sized companies, which were then not keeping abreast of developments or injecting their problems into the system. For the membership fee of £3300 (£1000 for small firms), members were offered state-of-the art reports focused on the technological problems identified by industry and voted for at meetings, prepared by the academic members. Several of these reports (Category 1) had been completed. The IPR belonged to FLOMIC. The proposed Category 2 projects were to be research projects which involved not only reviewing research in the open literature but also the pulling together of expertise in research organizations such as NEL in the UK and CERT in France and paid for by industry. However, only one project was completed, after considerable difficulty in persuading industry to fund it. A basic problem was the unwillingness to share intellectual property with competitors even though in principle the research was non-competitive and had little direct commercial value.

The structure and strategy of FLOMIC developed over time. It established the OPTRACA Sensors club. Members of OPTRACA got four state-of-the-art reports each year, and two pieces of experimental research funded by subscriptions to FLOMIC. FLOMIC was proposing to become a partner in a European Flow Measurements Standards group.

In September 1992, the steering group proposed to members at the AGM a mission statement and objectives which would 'Establish FLOMIC as *the* leading European Consortium of flow meter manufacturers, users and research organisations' (FLOMIC 1992). The proposals included:

- 'providing a forum and mechanism for technology transfer covering all aspects of flow measurement'
- 'support and enable "seed-corn" research for the development of related new technologies'
- 'promote the interests of the European flow measurement community – in Brussels'.

However, at that meeting, the members rejected the proposals. Instead it was decided to disband FLOMIC. One of its last acts was to hold a series of brain-storming sessions on the application of new and existing technologies to development of flow measurement instrumentation.

Fabrimetal

The Belgian trade association, Fabrimetal, is a professional body covering 1100 industrial companies in: metalworking, mechanical engineering, electrical engineering and electronics, transport equipment and plastics processing. It represents about 75 per cent of companies, but some small companies do not belong. Its function is more administrative than to do with technology transfer. Technical committees handle different subjects, for example field instruments. A big part of this subject is flow metering. Under this umbrella there are some contacts and some research that has developed between universities, users and manufacturers. There have been a number of reports including one on the effect of low conductivity on electro-magnetic (EM) meters. This was not an evaluation exercise but more of a research project to see whether the instrument could perform this function in a particular field. The process involved a literature survey and then field trials.

Euromet

In 1986, the UK, France and Belgium launched an initiative for European collaboration on measurement. This led to the EUROMET Memorandum on Understanding which came into force on 1 January 1988 (DTI 1989, 10). Membership of Euromet is open to national metrology institutes of EU and EFTA countries. The EU is also a member through the Community Bureau Reference (BCR) programme. Historically, the workplan has been developed principally from the advance of EUROMET and its equivalents. Since it was derived by recommendations from academics from laboratories, it contained projects that were academic in nature. Consequently, the vast majority of applications were from national laboratories (Brookes 1994b, 27). Three types of collaborative activity are envisaged through EUROMET: a) common measurement standards, b) facilities and c) research. EUROMET provides for the establishment of co-ordinated research programmes which may be undertaken at a single site or concurrently at a number of sites.

6.2   The electronic components industry

Industrial organizations have no significant role in creating links between member firms and the PSSB in this industry. The electronics component industry in the UK is represented by the Electronics Component Industry Federation (ECIF) and the Printed Circuit Board Federation (PCIF). In 1991, the ECIF had 120 members and the PCIF had 114 members. It has no interest in generating interaction with the PSSB. In France, GIIPRA (an amalgamation of the Federation des Industries Electriques and Electroniques and the Federation of des Industries Mechaniques et Transformatrices des Meaux plus 13 manufacturers associations) 'represents manufacturers of automatic machines and units, components and parts, specialised service as well

as engineering companies'. It has 165 members, of which 37 are electronic component companies. SYCEP and SITELESC are other organizations which represent the industry.

The international society for hybrid micro-electronics (ISHM) has as
one of its aims to bring universities and industry together. It has three
main chapters, ISHM US, Japan and Europe, with smaller local chapters. Its objective is to promote hybrid circuits on a scientific basis. Its
role is limited as it tends to be dominated by industry, and because the
market for hybrids is not increasing.

7.　Summary and discussion

This chapter identified a series of factors which potentially influence
the conditions under which firms engage in networking, collaboration
or externalization of innovation. At the beginning it was pointed out
that the PSSB may not be the obvious source of externally generated
technological input. Customers and users for many firms are the
most important sources of innovation. Moreover, horizontal linkages
with other firms might be more effective in contributing to in-house
innovation than links with the PSSB.

The second part of the chapter examined the characteristics of each
sector and showed that are more differences than similarities between
them. The most important difference is in the pace of technological
change. At the leading edge of the electronic components industry the
speed of change is due to engineering and scientific advances such as
in materials technology and by market opportunities in such areas as
telecoms, computing and aerospace. The pace of change has traditionally been far slower in flow measurement across the industry. This is
due both to a longer life of a particular design and conservative users.
In contrast users in electronic components are drivers of change. The
breakdown in the historical buyer-supplier relationship between users
and large firms in the flow measurement industry has allowed smaller,
innovative firms to assume more dominant positions in the market.

A second difference is in the type of competitive forces which drive
innovation. A general feature of the flow measurement industry is the
downward pressure on price of instrumentation caused by increasing
competition and low profit margins leading to lower investment in
R&D. Other factors were country-specific. For example, the UK
National Measurement system has enabled UK flow measurement firms
to compete on the basis of accreditation. The less well-developed
system in France and the absence of an equivalent in Belgium has

meant that in both countries individual firms rather than the domestic industries as a whole have adopted accuracy as a competitive measure. Other market structure factors which appear to contribute to the momentum of technological change in the UK's flow measurement industry are hostile take-overs, the privatization of gas and water supply, and the withdrawal of industrial users from meter design. The changing market structures resulting from process of privatization of utility companies have increased the choice which privatized firms can make about how they will compete in which markets.

A third major difference is that representation by trade associations appeared to be more significant in flow measurement than electronic components. However, the creation of FLOMIC was a result of key individuals identifying a need to tie firms rather than individuals into a system of innovation by institutionalizing membership of a formal innovation filiere.

The main similarities were that in some sub-markets innovation strategies were not those of continuous innovation; there were the opportunities for small innovative firms entering niche markets, particularly in the UK, and the frequency of take-overs. In electronic components, there were some similar pressures to those in the flow measurement industry which slowed the pace of technological change in some parts of the industry. These included declining markets, particularly defence, which meant that the technical priorities of other markets became more important, for example the demands of emerging markets such users in the automotive industry, and the downward pressure on prices for standard components.

Part Three
Comparative Analysis

5
Externalization Patterns in the Flow Measurement Industry

Introduction

This chapter discusses the main findings of interviews with firms in the flow measurement industry in the UK, France and Belgium. It takes a thematic approach comparing innovation strategies, patterns of linkages with universities and national laboratories. The objective is to identify the major factors influencing the extent and form of the sample's links with the PSSB. The chapter is in five sections. In the first, the sample is introduced. In the second, the firms' innovation strategies are identified. The third section examines the extent of externalization, networking and collaboration. In the fourth, the problems and benefits associated with interaction with the PSSB are discussed. The final section summarises the main findings.

1. The sample

Twenty-five firms were interviewed: 13 in the UK, five in France and seven in Belgium. Firms fell into four categories:

(i)   manufacturers of utility and industrial meters (15 firms);
(ii) research centres (4);
(iii) users (6 firms);
(iv) in Belgium, sales offices of multinational companies (4 firms).

The major suppliers of gas meters in each country were included in the sample. Meter manufacturers consisted of two types: those which make meters for utilities (gas and water), and industrial process meter manufacturers.

The sample comprised a mixture of independent firms and foreign-owned firms. The wave of take-overs in Europe throughout the 1980s and early 1990s engulfed most of this sample as nine (six UK, one French and three Belgian) had changed ownership between 1983 and 1992. The sites of manufacturing firms visited ranged in size from 15 to 110 employees. The technical departments of user firms were mainly small units employing fewer than 30 people. Details of date of formation, and main products/markets and size of site visited are presented in Table 5.1.

The UK sample consisted of nine meter manufacturers and four user firms, two of which were research centres in UK and will be discussed in the role of users. This is an example of a traditional industry which is firmly rooted in the UK by virtue of the particularities of its markets. The UK firms were generally older than those in France and Belgium. One could trace its ancestry back to the Napoleonic wars in the eighteenth century; another was founded in the 1920s. Most firms, both manufacturers and users, were UK-owned. Acquisition patterns were mixed. Some had been bought by overseas firms while two of the manufacturers had returned to domestic ownership from being part of US-based groups. The smallest firm had recently been acquired by a UK group. The headquarters of the four non-UK manufacturers were in the US, Denmark and Germany. The user firms were British Gas, a privatized water company, a petrochemical company and the chemicals division of large manufacturing firm. Markets were mainly domestic rather than export. Competition from German, Danish and Japanese firms, with their longer-term approach to innovation, was mentioned by a number of firms as threatening traditional markets.

France traditionally has been of less significance both in terms of production and innovation than the UK. This is reflected in the relative youthfulness of the industry, the presence of more independent firms and the more fragmented production/innovation system. The French sample comprised three meter manufacturers and two research centres. All but one had been formed since 1945. Two of the three manufacturers were independent, one French and one German. One formerly independent firm had recently been taken over by a US MNC. The French independent firm's founder was still in charge. When Gaz de France was formed in 1947, this firm was one of six suppliers of meters; the other five had since been acquired by the US based Schlumberger. The research centres were the flow measurement research laboratory of Gaz De France and the European R&D centre of a US instrumentation manufacturer, to which Firm 3 in the UK and Firm 2 in the Belgian samples belong.

Table 5.1 The flow measurement sample showing date of formation, employment and markets

Ownership	Date formed	Site employment (group employment)	Main products/markets
UK			
1. Denmark	1932	55 (14,000)	Manufacturer, electromagnetic flow meters/ water and chemical industries, 60 per cent sold to the UK
2. Germany	1953	140 (5000)	Manufacturer, mainly electro-magnetic gas flow meters, some vortex and coriolis meters for industrial control e.g. brewing, chemicals, petro-chemicals, and paper and pulp industries
3. USA	1920s	1000 (45,900)	Manufacturer, gas meters
4. UK	c1916	1000 (3,000)	Manufacturer, meters for food process industry
5. USA	1960*	450 (7,000)	Manufacturer, industrial meters
6. UK	1820s	1100 (1,114)	Manufacturer, gas meters, UK 75 per cent of its sales. Export to mainland Europe, the Far East and Middle East.
7. UK	1980	25 (3000)	85 per cent sales in UK, the rest elsewhere in Europe nuclear energy industry, sewage and landfill gases and automobile engine testing
8. UK	1940s	120 (15,000)	Manufacturer, supplies standardized fluid flow meters and process control systems for use in process industries e.g. petrochemical, food, pharmaceuticals.
9. UK	18th century	300 (600)	Manufacturer, water, power generation, chemicals and petro-chemicals industries
10. UK	not known	25 MNC	User – chemicals
11. UK	not known	240 8000	User – water supply
12. UK	not known	7 not known	User – gas supply (research centre)

Table 5.1   Continued

Ownership	Date formed	Site employment) (group employment)	Main products/markets
13. UK	not known	7 MNC	User – petrochemicals (research centre)
France			
14. Germany	1962	170	Water meters 15–20 per cent national market share
15. France	1930	160	Turbine and positive displacement gas meters for industrial markets, in niche markets: 50 per cent petroleum industries, 25 per cent aircraft industries, 25 per cent chemicals and special applications. Agency for a US company's products. Turnover in 1991 was FF 100 m. Fifth of output is exported, with major export areas in North Africa.
16. USA	n/a	500 (45,900)	Gas and water meters (research centre)
17. France	1948	170 (45,900)	Turnover in 1991 was FF 314 m of which 15–20 per cent was exported.
18. France	1947	12 (30,000)	Team involved in flow measurement for domestic and industrial applications, gas supply (user firm) (research centre)
Belgium			
19. UK	not known	14	Sales Two main lines 1. Open channel flow and 2. EM and ultrasonic meters. Six markets account for 90 per cent of the business – water treatment petrochemical, chemical industry, food, pharmaceutical and drinking water. Agency for other meter companies, e.g. German, French, Danish, Swedish.

Table 5.1 Continued

Ownership	Date formed	Site employment) (group employment)	Main products/markets
20. USA	not known	170 (45,900)	Manufacturer, gas and electricity meters 80–85 per cent of the gas meter market. Suppliers of the range of products produced by the group as a whole.
21. Belgium	1946	15	Manufacturer, differential pressure meters. Major customers are the sales operations of MNCs located in Belgium and domestic customers in petro-chemical and nuclear industries, chemical plants, petroleum steel companies, power plants, thermal and nuclear etc. other meter suppliers
22. USA	not known	100	Agency Markets: 90 per cent in Belgium, the rest France, Holland. Sell directly 70 per cent to end users, rest to other flow companies – importers. Turnover – about BF 50 m.
23. USA	not known	53	Sales, electro-magnetic flow meters, fluidic, vortex, mass flow and high level areal flow meters. Power plants, metal feral and non-feral, food and beverage, chemicals. Turnover BF 250 m 1991
24. France	1975*	30	Sales, water meters. 25 per cent of sales are in Belgium, some 60 per cent rest of Europe with other markets in North Africa (12 per cent) and a small amount in central America.
25. Belgium	not known	45,000	Chemicals (user firm)

Source: Author's survey
* date of acquisition by current owners

The marginal status of Belgium in the international flow measurement production filiere is reflected in the dominance of sales offices in the sample. Only two firms were meter manufacturers, one of which was an independent Belgian firm, four were sales offices of foreign MNCs based in the UK and the US. The seventh firm in the sample was a user firm, Belgium's largest company, a chemicals manufacturer.

2.   Innovation strategies

The case studies revealed considerable variations in innovation strategies. Differences arose from the characteristics of individual firms, particularly the impact of ownership, relationships with user firms, firms' use of national and EU regulatory programmes and, in the UK, the impact of privatization.

A major factor in the kind of innovation strategy adopted was the degree of autonomy over both the direction of R&D and the R&D budget. The independent companies were among the most innovative. The subsidiaries of the independent German manufacturers based in the UK and France were both allowed considerable freedom in deciding their innovation strategies. For the one of the German firms based in the UK, a change of emphasis within the parent company had meant an increasing focus on metering, requiring a rapid process of catching up on meter technology. It was one of the most innovative of the UK sample. However, being innovative does not necessarily imply that R&D effort is directed towards radical changes in technology. The firm saw its competitive advantage in providing a good service to customers. This meant that engineering developments were directed towards reducing the price of meters. The UK operation of the independent Danish company was reputed to produce some of the best meters in the market but in this case technical improvements were made outside the UK in its R&D centre in Denmark. FLOMIC membership was through the parent company rather than the UK operation. In Belgium only the independent company was undertaking R&D in meter design.

In contrast those which belonged to US MNCs in all three countries had far less control over R&D activities. One of the UK-based operations of a US group was limited in its freedom to innovate because R&D strategies of all the companies in the group had to be approved by R&D headquarters in Paris (included in the sample, Firm 16). Moreover, its R&D department has suffered under-funding under previous ownership because flow measurement was not a core business. The response to the call for new gas meter designs by both British Gas was based in the research centre in Paris rather than in the UK manufacturing company. The other supplier of meters to British Gas, a UK firm, had failed to respond to the challenge and was one of the least innovative companies. A second US-owned operation based in the UK had reduced the level of R&D in recent years.

Relationship between manufacturers and users are a fundamental component of the innovation filiere in the flow measurement industry. However, they do not always have positive effects on innovation. Innovation in some markets has traditionally been stifled by a lack of interest in new meter technologies on the part of user firms. For example a UK manufacturer of meters used in the food industry had limited opportunities for innovation because of the conservative attitude of the industry. The Belgian user firm, the leading Belgian chemical company, had never been involved in meter design and therefore had not developed technical relationships with meter manufacturers. It was primarily interested in the performance of meters which it monitored through membership of the WIB. Other users in Belgium were demanding improved meters which used less energy. This development did not require major advances in technology. The programme of modernization of the water treatment industry in Belgium had created new market opportunities but for existing rather than new types of meter. On the other hand, users of industrial meters in France had began to demand greater accuracy. This gave one manufacturer the incentive to invest in the development of new techniques.

The impact of regulatory change is most evident in the UK where privatization of water companies and British Gas had a major impact on R&D strategies. The water company in the sample used the opportunity provided by privatization to move into the design of water meters as a means of satisfying its customers' service requirements.

Several firms in the UK mentioned EU legislation as being a driving force for innovation. The smallest UK firm expected that new designs to meet EU standards would mean increased market opportunities. Another cited EU legislation on effluent and smoke emissions as requiring improved meter design.

3. Externalization, networks and collaboration

3.1 Extent of linkages

The norm in this industry is for firms to have some kind of link with universities and/or national laboratories. The form of link ranged from

Table 5.2   The flow measuring industry sample: links with the PSSB and government support

	Sample size	University link number (per cent)[*]	National laboratories link number (per cent)	National funding number (per cent)	EU funding number (per cent)
UK	13	12	1	3	0
		(92)	(8)	(23)	(0)
France	5	3	1	2	1
		(60)	(20)	(40)	(20)
Belgium	7	1	0	1	0
		(14)	(0)	(14)	(0)
Total	25	16	2	6	1
		(64)	(8)	(24)	(4)

Source: Author's Survey
[*] Percentage of national sample

formal research collaborations, to problem solving contract research, through students undertaking projects within the company to informal contact with individuals. The number and proportion of firms in each country with an interaction with one or more institutions in the PSSB is shown in Table 5.2. The table shows the links were most frequent in the UK and least in Belgium. Several firms, two in the UK, two in France and six in Belgium, had no contact with the PSSB.

3.2   Funding

The general pattern was for firms to fund research in the PSSB directly. This is illustrated in Table 5.3. On the whole, there was a low success rate in obtaining funding from national or EU programmes. Only a quarter of all firms had funding for collaborative research from their national/regional governments and only one, the Paris-based research centre, had an award under an EU programme. Few UK and French firms participated in co-funded studentships schemes. None of the Belgian firms had studentships.

In the UK, one CASE studentship could not be taken up even though the project was approved and the student was selected because SERC had run out of money. The student could not wait another year for the money come through. The company decided that it was not the right

Table 5.3   The UK flow measuring industry sample: forms of links with universities and national laboratories

Ownership and location	University/National laboratory links	Notes
1. Denmark (South West)	No formal research links	Informal links and attendance at seminars and courses at Cranfield, calibration and testing at NEL
2. Germany (North West)	Cranfield, NEL, FLOMIC	CASE studentship, plus research contract
3. USA (North West)	Sheffield, Liverpool, Cranfield, Manchester, NEL and through FLOMIC	Respondent was first chairman of FLOMIC Contacts with NEL mainly for testing
4. UK (South West)	Liverpool NEL Liaison group	LINK
5. USA (South East)	Now few formal links	Links through Optical Sensors Club (OSCAR) and FLOMIC
6. UK London	No research links	Student placements
7. UK (South East)	Surrey FLOMIC	Industrial placement scheme
8. UK (South East)	Oxford	Three year LINK project
9. UK (South East)	No links	Ex-FLOMIC members
10. UK (North West)	Oxford	Three year LINK project
11. UK (West Midlands)	Cranfield Warwick, City, NEL	Four year contracts
12. UK (London)	Cranfield, Erlangen	Problem solving
13. UK (London)	Main contact through FLOMIC, respondent FLOMIC committee member	Informal contacts with individual academics

France

1. Germany (Rhône-Alpes)	No links	Looking, has had links in the past
2. France (Paris)	Poitiers, and EXA-DEBIT	Student sponsorship for 15 months supported under CIFRE scheme

Table 5.3   Continued

Ownership and location	University/National Laboratory Links	Notes
3. USA (Paris)	Paris IV, Ecole Polytechnique, Ecole Superior de Physique et Chemie, Orsay, EISEE, Grenoble, Cambridge, Surrey, Frauhofer Institute (Munich), Centre National Microelectronics (Spain).	Projects funded by the Ministry of Research and Technology and Defence Ministry. Two current projects. Regular grants from the Ministry of Defence. No assistance from regionally organised agencies such as ANVAR because the group was too big to be eligible for grants. ESPRIT consortiums.
4. France (Paris)	No research links	Student placement, testing and validation in local universities some 20 students from engineering schools working in the company in the vacations – means of filtering new employees.
5. France (Paris)	Bordeaux, Paris VI, CERT, Poitiers, FLOMIC, Surrey	Contacts with CETIAT in Paris, technical centre. Long term collaboration with CERT.

Belgium

Ownership and location	University/National Laboratory Links	Notes
1. UK (Brussels)	No links	
2. USA (Brussels)	No links	
3. Belgium (Wallonia)	Louvain-la-Neuve, Liege	One year project on software development for meter
4. USA (Flanders)	No links	
5. USA (Flanders)	No research links	Company donates equipment and provides information for the education of students
6. France (Wallonia)	No research links	Testing, adaptation and validation of meters and informal links with Liege
7. Belgium (Brussels)		

time for spending another £10,000 each year for three years to fund the student directly. The French manufacturer's CIFRE studentship was with the University of Poitiers. A project was agreed in 1992 which would build on technological advances in the firm. The student was undertaking a theoretical study of interference and effect on accuracy of measurement. It was orientated to a fundamental advance in controlling the performance of instruments.

The low take-up of public funding is in sharp contrast with the electronics industry (see Chapter 6) and is an indicator of the kinds of 'information' and 'competence' gaps in both individual firms and collectively in the industry. In this case firms did either not know how to apply for funds, or were unsuccessful in their applications. The Belgian firm with an innovation support grant from the Wallonian government to support the research in the University of Louvain-le-Neuve stressed its importance. The addition of this external expertise reduced the timescale of the research effort, enabling the company to get the product to market sooner than if it had been necessary to fund all the development itself.

The experience of the UK sample towards obtaining public funding fell into four categories:

(i)   those which had applications funded. Three firms had LINK projects. The first was between Firm 4 and a food manufacturer; the second was between Firms 8 and 10 and one of the universities in the sample;
(ii) a firm (Firm 3) which had been unsuccessful in its application;
(iii) a firm which had no information about how to apply (Firm 7);
(iv) a firm which had been discouraged by the amount of effort involved (Firm 11).

Firms in all four of the categories identified above complained about the UK system of funding industrial innovation. There were three major complaints. The first was that the UK does not support near market research; the second that the complexity of LINK rules inhibits interaction between universities and industry at the 'pre-competitive' level; and the third was that the DTI did not recognize the needs of both industry and the science base, and generally did not do enough to support this industry. Earlier schemes such MAPCON and Support for Innovation were preferred to LINK because they provided better ways of facilitating improvements in firms' technological performance. Several suggested that the DTI could provide better support for entry to European schemes.

The view of the JIMS Co-ordinator, also interviewed, provided a contrasting point of view on LINK funding. He said that even though the flow measuring industry is only a small part of the instrumentation industry, its proportional representational interest 'would be greater and its voice would be heard if they shouted loud enough'. He indicated that if any consortium came along with new ideas about the measurement of fluids, gas and so on, the DTI would be very interested in exploring it further. He had mainly acted in a responsive mode because there had not been the time to do otherwise. He felt that there needed to be more representation on standards committees.

Similar comments to those made about the difficulties in obtaining LINK awards were made about EC funding. Two UK firms (Firms 2 and 5) had failed applications, and another had been deterred by the amount of effort involved. In one case failure had major repercussions. Prior to take-over, Firm 2 had applied for a £7 million Eureka project involving universities in the UK, Netherlands and Norway. The research would have covered all aspects of fluidics. The respondent believed that if the application had been successful, acquisition might have been prevented.

3.2   Geographical pattern of linkages

The general pattern was for firms to have links with universities and national laboratories inside their own country and for these to be local, defined as being institutions within 50 miles (see Table 5.3). Only three firms, British Gas and the two research centres in France, had direct contacts with foreign universities and national laboratories. Two others had indirect links with non-UK universities. These were one in France and one in Belgium which had contacts with universities in neighbouring countries through sister companies.

The number of universities involved with the UK sample was small. Only nine universities were mentioned by the seven firms which had formal interactions. In most cases contact was with only one or two universities. British Gas had links outside the UK through the Siemens research centre in Germany.

In France, the pattern was that the manufacturers had a limited range of links which were usually with local universities. The two research centres had both the highest number of links and the most extensive range of contacts, inside and outside France. Both had links with Surrey University in the UK. Only the flow measurement research laboratory of Gaz de France had close contact with CERT. The Belgian manufacturer's links were with two local universities.

The reasons why firms chose to work with the PSSB and what the forms of assistance they pay for or receive through informal means are considered next. The lists of possible answers was taken from the work by Charles and Howells (1992). As in their study, firms could give as many responses as were appropriate.

3.3 Motivation

To the question of what motivated firms to seek external expertise from the PSSB, the two most frequent answers were inadequate scale of internal R&D and access to new areas of expertise (Table 5.4). These were followed in importance by access to existing information, and

Table 5.4 Motivation for contact with the PSSB

Motives for collaboration	UK	France	Belgium	Total
Universities				
Cost saving	6	0	0	6
Inadequate scale of internal R&D	7	5	1	13
Standards development	3	0	0	3
Access to new area of expertise	8	3	0	11
Access to existing information	4	3	0	7
Strategic benefit/opportunity	5	0	0	5
Participation in public programmes	5	1	0	6
Access to public sector markets	3	0	0	3
Complementary expertise	0	1	0	1
Internal problem which could not be solved	6	1	0	7
Other	2	0	0	2
None	1	1	6	8
National laboratories				
Cost saving	3	0	0	3
Inadequate scale of internal R&D	4	3	0	7
Standards development	7	0	0	7
Access to new area of expertise	4	3	0	7
Access to existing information	3	0	0	7
Strategic benefit/opportunity	3	0	0	3
Participation in public programmes	3	1	0	4
Access to public sector markets	3	0	0	3
Complementary expertise	0	1	0	1
Internal problem which could not be solved	5	1	0	6
Other	0	0	0	0
None	12	2	7	21

internal problems which could not be solved. Other factors were access to existing information, participation in public programmes and cost saving. These reasons were common to firms in all three countries. The findings indicate both the general importance of research in the PSSB to the companies and the limitations of in-house resources.

Some motivating factors, however, were specific to the UK or the UK sample gave the majority of the responses. These were cost saving, strategic benefit/opportunity, standards development, and participation in public programmes. Cost saving was mentioned by half with regard to universities and a quarter to national laboratories. In contrast with the firms studied by Charles and Howells (1992, 148), there were economic benefits from contracting R&D activities to the PSSB for this sample. Externalization provided the opportunity for technical developments to be made which could not or would not have been undertaken in-house.

The national laboratories did not have the same function within the innovation filiere of the UK as in France. The UK sample's links with national laboratories tended to be more applied than those with universities. Seven firms had contacts with NEL and three with Harwell. Most UK firms had some interaction with NEL because of its calibration and testing function and for the development of standards as firms geared towards meeting ISO 9000. In contrast, three of the French sample were motivated by the need to gain additional support for in-house activities.

3.4 Forms of assistance

The most frequently mentioned form of assistance sought from universities in all three countries was basic scientific knowledge (13 firms) (Table 5.5). This was far more frequent than new generic technology development (six firms). Again there was a difference between the UK and the other two countries in the kinds of assistance sought. Only UK universities were involved in solving product development problems. This possibly reflects the stasis in the industry in the UK as incremental change rather than radical developments, except in the case of gas meters, is the norm.

National laboratories in the UK generally provided help with routine testing and validation. Only the water company looked to NEL for new generic technology. Technology transfer functions were met by NEL's flow liaison group to which several firms belonged. The relationship between CERT and Gaz de France was that of research collaborators. They had been working jointly for one to two years on a problem

Table 5.5   Forms of assistance sought from universities and national laboratories by the flow measurement sample

Forms of assistance	UK	France	Belgium	Total
Universities				
Basic scientific knowledge	8	4	1	13
New generic technology development	4	2	0	6
Assistance in new product development	1	2	1	4
Developing testing routines	2	0	0	2
Solving product development problems	3	0	3	6
Improving production processes	0	1	1	2
Testing or validation services	2	0	1	3
Adaptation of technology to local conditions	0	0	0	0
Other	1	0	0	1
None sought	2	0	5	7
National laboratories				
Basic Scientific knowledge	0	1	0	1
New generic technology development	1	0	0	1
Assistance in new product development	0	0	0	0
Developing testing routines	5	0	0	5
Solving product development problems	0	0	0	0
Improving production processes	1	0	0	1
Routine testing or validation services	6	4	0	10
Adaptation of technology to local conditions	1	0	0	1
Other	0	0	0	0
None sought	4	1	7	12

Source: Author's Survey

related to accuracy of readings in pipe positions. The relationship had been built up over a number of years.

In France, there were varying degrees of awareness and methods of keeping informed of developments in universities. Some years earlier Gaz de France had looked for research expertise for ultrasonic meters for answers to theoretical problems associated with the failure of meter systems to work properly. To achieve this aim, it had funded a two-year project with the University of Bordeaux, which has a laboratory specializing in ultrasonic technologies; it had taken on a PhD student for three years from Paris VI University; and was working CERT on standards development and with EXA-DEBIT on testing and validation.

The European R&D centre of the US MNC was funding universities for basic research, while working on prototype development in-house.

In general the centre adopted the strategy of being state-of-the-art and maintaining this position by funding research in universities. However, even this firm did not always compete by being first with new technologies, instead it sometimes chose a 'fast second' approach. Universities were also a source of recruitment and personal contacts. Both were very significant in determining which universities would be funded to undertake research projects.

3.5   Duration and changes in patterns of linkages

Firms were questioned on the changes in their relationships with the PSSB, how long the links had lasted and how they were funded. The pattern was mixed. About half of the firms in the UK and France and one in Belgium had made medium- to long-term commitments. For the rest, contact was for shorter periods of time. In the UK, four projects involving six of the sample lasted for up to three years. In France the two research centres plus a manufacturing firm had long-term relationships with universities and/or national laboratories.

Trends in the relationship between industrial innovation and industry/PSSB links differed between the countries. In the UK there was evidence of both a decline in interaction and of new long-term commitments. Five firms had decreased the level of their direct contact with universities over the previous five years. In three firms this was either a direct consequence of acquisition because efforts had been redirected towards product improvement and new product development rather than to more speculative activities. One of the three was a small company which had been forced to cut back on R&D following acquisition by a large UK group. One of the larger firms had substituted independent direct links for funding projects through FLOMIC and OSCAR and looked to industry research organizations such as SIRA to keep in touch with the latest developments. A fourth firm, one of the users, was formerly active in working with manufacturers on meter design but the recession of the late 1980s had meant that the engineering budget had been cut. This limited the resources available for working on the development of a coriolis meter. However, in spite of formal interaction decreasing, the individuals formerly responsible for collaborative activities continued to monitor research in universities.

In France and Belgium there was some evidence of an increase in interaction with universities albeit from a small base. One of the manufacturing firms in France had strengthened its links following the appointment of a new engineering manager. He was to be responsible for the search for a university which it could fund to undertake research which involved the identification of different physical principles currently in use in meters with the objective of designing a new meter. He would be looking to CRITT to subsidize any research with a university when a suitable one was found. The European R&D Centre had recently become more involved in university collaboration, particularly in the area of micro-sensors. Its interest in interaction was helped by changes in the science base – CNRS laboratories had become much easier to work with since they became more interested in working with industry in the late 1980s. There was little prospect of an increase in the general level of collaboration in Belgium because firms were either sales offices or, in the case of manufacturing plants, R&D was undertaken outside Belgium.

In the UK, longer-term projects fell into two categories, those directly funded by firms and those under LINK programmes. The former included one between the independent German firm and a university, and the water company which was funding research at three universities and at NEL. For this firm, the projects were part of the process of getting 'up to speed' on developments in metering. Two projects organized under the LINK Programme each involved a manufacturer, user and a university.

3.6 Representation of interests

A key element in the organization of innovation filieres is the industry's capacity to exert control and create a coherent voice. One means by which this is possible is through representation on committees responsible for the funding of research. The UK flow measuring industry had few representatives on research council committees. Only the Director of R&D of Firm 5 served on an academic committee. This was the advisory board for a university department's work on sensing technology. He also chaired the sensor and transducer group at GAMBICA. For him, these positions were a deliberate policy to keep in touch with experts in the field in university departments. Three companies had representatives on ISO standards committees.

3.7 FLOMIC and trade associations

The importance of FLOMIC in creating and sustaining networks cannot be underestimated in this industry. It played a major role in integrating academic research into industrial innovation activities. Most of the UK firms and one from France were members of FLOMIC. Another had been intending to join while another had left.

The general impact of FLOMIC was that the level of interaction between industry and the PSSB in the UK increased. For some firms the organization became the main formalized source of information about academic research. Some of these previously had had direct links with universities while for others, such as the water company, it was an important source of new contacts. There was general agreement in the sample that FLOMIC worked well as a talking shop and as a dating agency. The state-of-the-art reports were universally commended. For one firm they shaped thinking in both in marketing and R&D departments. Membership had advantages for two of the smaller companies. It brought status with other members and was a means of getting support for R&D projects within the company. Gaz De France had been involved with UK research through FLOMIC since the late 1980s. Its Chief Engineer's links with the UK arose from the need to keep abreast of new developments. FLOMIC provided a source of contacts. The respondent described the UK as, 'the ancient land of flow measurement'.

At the same time, most firms were critical of FLOMIC in a number of respects. It was not perceived to be the answer to many of the industry's problems. The most common complaints were that the interests of user members should have had priority and that sub-groups dealing with the needs of particular industry sectors should have been set up. The reality was that users were either increasingly reluctant to participate or found it difficult to justify membership to their boards of directors. Other firms commented that the orientation was too academic and driven by that side of the consortium; membership was not worth the money. This was because of necessity, projects were too generalized due to the range of interests which had to be satisfied, because the results were widely available and that FLOMIC attached strings to work that it funded. It was able to do because it owned the IP.

3.8   Benefits from interaction

Several kinds of benefit from interaction with universities and national laboratories were found in the survey. The most common was help with product development. This ranged from product design, as in the Belgian case where the academic was involved with software development, identification of theoretical basis for the meter, as in France, and hands-on development as in the case of meter under the second of the two LINK projects.

A second major benefit was that of education. In the UK, universities were providing the extra service of educating industrialists who had previously not worked in flow measurement, who were not up to date with new developments, or who wished to keep up to date with developments. In the UK two manufacturers and a user firm said that working with universities overcame the problems of a severe information gap in the principles of metering. In the case of the German firm, expertise in flow measurement had been built up by two employees undertaking studies at Cranfield University and the recruitment of engineers with degrees in relevant disciplines. Engineers in the water company were also brought up to speed in metering technologies by developing links with universities. Although the Danish firm had no formal links with universities, developments were monitored through journals, and through information obtained by the senior engineer, who did not have a background in flow measurement, attending seminars and conferences. These contacts made it possible for him to learn about basic principles and theoretical problems related to electromagnetic meters. The contacts between Gaz de France and the UK flow measurement innovation system in general fulfilled the same function. Access to important information on standards, as was the case in Belgium, was another education benefit.

A third advantage of interaction reflected the low status of engineering projects inside some UK firms. The association with universities provided legitimacy to the development. For one the strategic decision to fund research was going to influence the way the business would develop. There was also the benefit of bringing market acceptance to customers. Two UK and one Belgian firm said that the university collaborations enhanced the image of the company. The Belgian firm had highlighted this project in its publicity material.

Recruitment was not mentioned as a major benefit. In this industry industrial and academic careers are separate. Only the French R&D centre cited the possibility of students being a source of recruitment, while for one UK firm, short-term problem solving by universities meant that company did not have to recruit engineers.

The LINK projects were important to the firms involved in several respects. The project involving Firm 4 was worth £250,000. It was intended to improve the design of a meter already in development. LINK funding was crucial as it had inflated the importance of the project to the firms' accountants, which enabled the technical department to get more resources devoted to the research. Some progress had been made. The university had completed the research, the manufacturer had finished the product engineering and had built prototypes, and the user had been involved in the marketing and interface with the customers. At the time of the study, however, there was a danger that it might not be completed because the firm had run out of money. The project had been run initially for three years and was then extended for another year because of delays caused by the firm. The project was far from the market, and was not as high a priority as other, more commercial projects. Even so, the project had been sufficiently successful to satisfy internal assessment criteria for the firm to look for another at the end of the current LINK project. The investment in new technology could be licensed to other firms and further links would mean a new product and supply of services to go with it. The project had led to a much improved market position. It had also achieved leverage on the money spent by the company, provided access to good brains in the university, and the results of in-house research had come earlier as a result of the interaction. The company felt that it would be able to develop more ideas if it were easier to secure government funding.

LINK funding was the decisive factor in whether a project would be approved for one firm. In the project involving Firms 8 and 10, the work would not have been undertaken if Firm 8 had had to fund all of its share of development costs. This was one of the first JIMS projects. It started in 1986. Earlier applications to the DTI for funding had failed. The partners had one key advantage when submitting this application which was that the user firm's research and engineering associate chaired one of the LINK schemes. This meant that the rules of the game were known. The project was initially for three years. The total cost of the project was £400,000 of which the user firm contributed £90,000.

3.9   Problems

The most obvious difference between the UK on the one hand, and France and Belgium on the other, was that of the UK-specific problem of allocation of IPR. In France the arrangement in one collaboration was that the university held the theoretical IPR and the firm the commercial rights. The problems experienced by Gaz de France were to do with the lack of breadth of research in French universities and national laboratories. Problems reported in the Belgian case were related to some differences in approach but were not enough to hinder the conduct of the project.

Half the UK sample (six firms) had difficulty in dealing with universities over IPR. Further problems were related to patents and publications. The general tendency was for firms to demand exclusive rights and for universities to try to retain them. For example, Firm 2 was frustrated by a university insistence on retaining the rights to work which underpinned a FLOMIC project although it would release the rights to the specific work undertaken on the project. The firm had hoped that this work might provide a business opportunity. The problem was made more acute in this industry than in others by the slow pace of technological change: IPR are valuable for longer in mature industries such as this than in industries characterized by a rapid rate of technological change.

The UK sample reported a series of failed relationships dating back over ten years. The main problems were to do with limited expertise in university departments and a lack of large-scale facilities such as steam rigs. In one case, a project with a nearby university had failed. A member of the firm's staff was to undertake a doctorate at the university but withdrew when the department's academic background was found not to be as promised. This soured future relations. The project then was moved to another university and a member of staff was enrolled as a part-time MPhil student to work. A second project followed. The firm ascribed the success of the project as being due to the department being 'set up for a short-term response' and having an appreciation of the company's point of view. It has also the technical infrastructure including a testing rig. The company had widened its links with this university, paying for an employee to do a PhD part-time under a CASE award, and were paying the department for academic support. Even in a good relationship such as this there was a fundamental difficulty over ownership of IP. The firm felt that it was constrained by not being able to have access to all available information in an area. It was further frustrated by universities' desire to extend what they are doing rather than generate new ideas. It found the benefits of interaction were that they reduced the need for hard research being done exclusively for the company.

One firm which had cut back on links felt that university researchers failed to recognize the limits to which industry works. In one case basic results from a project had been given to the company but the department had not produced the final report. The issue for the firm was to do with the communication of the findings, not the quality of research. Failure to produce the results of the research had caused a delay in in-house development. At the time of interview the final report was three years overdue. The firm was refusing to pay the £15,000 owed for research until the report arrived.

A further difficulty in the relationship is related to initiation. The industrialists felt that academics could do more to contact industry and identify where there could be a match of interests.

The restructuring of the national laboratories had so far not worked to the advantage of the firms in the sample. The danger expressed by some was that by becoming more short-term orientated, national laboratories would not be a longer-term research resource. All criticized NEL for being too slow and expensive. They were described by one respondent as being 'commercially immature'. One firm had cancelled a contract after four years, because the work was not completed. In spite of the cancellation, NEL continued the work and the report, described as excellent, was delivered at the end of the fifth year.

4. Five case studies

Case studies of two UK gas meter manufacturers, British Gas, a privatized UK water company, and a small independent Belgian company illustrate some of the ways that firms approach innovation and positive and negative attitudes towards the PSSB.

The first meter manufacturer (Firm 3) had for many years maintained extensive links with universities. It had involved five universities in in-house research into fluidics through direct funding over an eight year period and had contact with others through membership of FLOMIC. Recently, following take-over by a US firm, the company's innovation strategy had become more short-term orientated. The consequence was that universities were used for short-term projects when the R&D department had run out of time or in-house resources rather than for long-term developments.

Firm 6 had also reduced its level of interaction with universities in recent years and instead relied on in-house R&D and membership of OSCAR for technical information. The decrease in interaction was due to general company policy and to difficulties during previous attempts to develop links. Five years earlier the firm had approached a university with a view to funding a project on a thermal meter which a student was working on as part of a graduate degree programme. However, the work did not go ahead as the company decided that the work was too expensive and would have taken too long. Later it approached another university for the purpose of buying a meter design. The contact petered out because the university had moved on to a bigger project and did not respond to the company's request for a firm specification. The respondent's experiences led him to the view that universities should be more pro-active in finding out what firms needed and should appreciate that industry has budgetary constraints. Some contacts with universities had been productive. Some years earlier, two fourth-year students from the mechanical engineering department of a nearby polytechnic had worked for six months in the company on a major project. This had been the most successful form of interaction to date.

The research centre of British Gas had links with three universities in the flow measurement field either currently or in the recent past. With one it was supporting work on ultrasonics and with another it had supported a studentship. The general experience had not been positive. The respondent's view was that academics had not been active in informing the company about research activities. Moreover, the cost of research in universities was rapidly becoming more expensive with rising levels of overheads. This viewpoint was reflected in the strategy adopted for the new meter design. At the initial presentation stage, universities were not invited to the meeting. British Gas argued that their experience has been that flow measurement has been 'the Cinderella' of academic research and that there was no appropriate research to call on in either the universities or in NEL. This arguably represents a missed opportunity to bring universities and industry together at the outset in a major new initiative. This position can be supported by British Gas's subsequent involvement with two universities, one in the UK and one in Germany, who were brought in to assist in development problems 'where there are gaps in understanding' (personal communication). British Gas later awarded two university research scholarships, one at Cranfield and one at Newcastle, which commenced in October 1992.

The next two case studies provide a contrast with the three previous ones in that they are both successful collaborations. The first is an example of a new entrant to the UK industry, a newly privatized water company, which has been referred to earlier in the text. The second is from Belgium and shows how an *ad hoc* marketing strategy led to a research project and to the longer-term benefits of access to the science base.

In the first case, the company's most significant involvement with universities was a five-year contract with Cranfield university to design and develop a new flow meter device. The work could not be done in-house. The unit found it difficult to find universities with the mix of expertise to take things through to a real product. This university was chosen because it was at the leading edge in a number of projects in flow metering. Cranfield had the advantage of being able to go a long way towards pre-production prototyping. This enabled the firm to gain maximum benefit in terms of return. However, even this department was under-resourced for what the R&D unit would like to do. The practical difficulties of managing the project persuaded the unit that joint projects involving more than one university should be avoided and that it would be better to focus attention on developing one-to-one relationships. The firm felt that universities in general must start to take account of the fact that larger industrial concerns now search for technologies across the globe and cannot therefore expect that their only competitors are in the UK.

The recent relationship between the Belgian firm and a university arose out of an intensification of the firm's in-house innovation strategy. One component of this was its efforts to educate both its customers and local education institutions. It did this by way of a mailshot. The message to its customers was the necessity of conformity to standards and the technical advances which the company were making. The firm sought to inform the staff in further and higher education schools about the importance of flow measurement as a subject of academic research. A professor in the Thermal Dynamic department at the University of Louvain-la-Neuve found some errors in the technical details but thought that the idea was very interesting. He contacted the company and proposed a collaboration. As a result, the professor was paid on a one year contract to develop software. There was an additional advantage to the firm in that the professor was also a member of the ISO technical committee and could advise on how to make meters conform to the standard.

5.   Conclusions

The first general hypothesis was that there would an increase in links between firms and the science base. This is not the general pattern in the UK but was the case in some parts of the industry, in France and in one example in Belgium. Long-term links between manufacturers and universities and national laboratories were the norm for about half the sample, some of which were research centres in UK and France. The difference in participation in networks, collaboration and externalization between the firms in particular and between the different countries in general arose mainly from the effects of ownership and location of power.

What varied between firms was the extent to which individuals were working within the system. For research links to be maintained with the PSSB, there had to be a product champion within the company, who identified the key scientists and engineers, and who drove projects through. In some companies this person was working on their own but on the company's behalf because of their belief in the commercial value of innovation. The need for *personal* commitment in this industry arose for two reasons: (a) expensive research activity was squeezed as companies fight declining market shares, and (b) an absence of a political/institutional/industrial framework in which research links were the norm. The case studies show that personal commitment on the part of key individuals was critical in the development of research links between different institutions in the innovation process. The question is whether the existing links would be sustained if these individuals were to leave, or whether the concept of externalization had become embedded in the system. Moreover, it is doubtful given the current climate whether individuals in firms would now have the same freedom as a decade or so ago.

A major difference between the UK and France, and Belgium was the extent to which meters used were based on domestically engineered technology. In Belgium, the market was supplied by meters developed in the US, UK and France, whereas in the UK and France meters used were more likely to have been designed and built in the home country. This pattern was responsible for national variations in the demand for technology from universities and national laboratories, being much more prevalent in the UK and France than in Belgium. In France, however, the demand was for research into basic principles whereas in the UK and Belgium it was for more applied linkages. The tendency for the majority of the UK sample was to maintain low level linkages such as testing, courses and seminars as a means of monitoring developments rather than using them as part of a portfolio of innovation activities including recruitment, and indeed driving innovation. Applied upstream contacts were mainly at a medium level of contact for problems solving. This was a pattern of substitution of university research for internal R&D, associated with a decline in genuinely innovative activity over the last 15 years in a number of large companies particularly with US and UK parents. Where the strategy once was to put technologically advanced products on to the market in order to gain a lead, now the emphasis was on consolidation and standardization. In spite of some difficult experiences, industry would have preferred academics to be more proactive in identifying where there might be common interests. FLOMIC reinforced and re-focused the system of interaction based on product development.

The withdrawal of users from meter design was a problem specific to the UK. The effect had been to reduce the number of 'visionaries' in large organizations and deter entrepreneurial activities. Moreover, a consequence of the failure of UK users, for example, to adopt new technologies, is that they were being exploited in the USA and Japan, and not even Europe. It also meant that as the manufacturers were bearing the full costs of R&D there were fewer resources available for collaborative projects with the PSSB. The innovation filiere was being weakened because of the decrease in range of interactions which previously tied the different organizations (firms, users, universities and national laboratories).

6

The Views of Universities and National Laboratories: the Flow Measurement Industry

Introduction

In Chapter 5 it was shown that the main drivers of innovation in flow measurement were improving accuracy, new meter design and meeting competition by widening the range of meters offered rather than by radical advances in technology. Here the impact of those drivers on firms in the industry's relationship with universities and national laboratories is examined by reporting on the views of academics, scientists and engineers interviewed in each of the three countries. An objective is to examine the extent to which there is a match between the geographical and technological spaces in which individuals working in the PSSB and those in industry inhabit.

The chapter is divided into three sections. In the first, the characteristics of the sample are introduced: size of research groups, specializations, sources of research funding, the geographical scale of links and major changes affecting research priorities over time. The focus in the second section is the experiences of relationships with industry. The final section considers the implications of the survey.

1. The sample

1.1 Sample profile

The sample comprised 12 departments: seven in the UK, three in France and two in Belgium (Table 6.1). Three of the sample were national laboratories, two in the UK and one in France. In the UK and France, expertise ranged from specialisms in steam, multiphase flows, thermal

Table 6.1 Sample PSSB institutions with research in flow measurement in the UK, France and Belgium

Department	Staff	Specializations	Research funding
UK			
1. Fluid Engineering and Instrumentation (School of Mechanical Engineering)	3.5 academics plus 6 research staff	Electromagnetic and ultrasonic	UK industry
2. Thermal-Fluid Division (School of Mechanical Engineering)	6 academics, 15 RAs (two working on flow)	Multi-phase flow (2 and 3 phase flow), large steam facility,	FLOMIC, DTI
3. Engineering Science	Group led by senior academic, 1 JRF, 2 doctoral students	Sensor validation	LINK, SERC, studentships
4. Fluid mechanics group (Department of Mechanical Engineering)	6 plus 2 RAs, 3/4 research assistant, 3/4 UK funded graduate students, two post-grads on secondment from Gaz De France and other students from Japan.	Steam, flow measurement in pipes, turbine meters	FLOMIC, DTI, SERC studentships Contact mainly with large UK companies, EU small grant
5. Electrical Engineering	1 plus 4 RAs	Process tomography	SERC
6. Flow centre, NEL	40 staff	Fluid flow in pipes	DTI, UK, other European, US and Japanese industry, 10 per cent from overseas industry, BCR programme
7. Instrumentation Division, Harwell Laboratory	10	Instrumentation	UK industry, DTI
France			
8. Thermal dynamics group, (Joint CNRS/ University funded department physics department)	4	Thermal dynamics, particularly thermo dynamic instability	Industry, contracts plus CIFRE scheme CNRS

Table 6.1   Continued

Department	Staff	Specializations	Research funding
9. CERT	Flow measurement group 2 senior staff, one visiting professor, 1/2 graduate student, and 1 student from Surrey University.	Basic studies of dynamic behaviours of different types of flow meters in disturbed conditions, downstream	90 per cent industry, 10 per cent government 80 per cent research income in 1991 from Gaz de France.
10. EXA-DEBIT	n/a	Applied metrology	Industrial membership
Belgium			
11. Thermal dynamics within thermal engineering faculty	4.5 25–30 (15 per cent funded on research contracts)	Thermal dynamics	Belgian, French and other foreign Industry
12. Agricultural sciences Applied Sciences faculty		Fluid dynamics	Belgian industry

dynamics, electronic engineering and software development. Most flow specialists in universities worked in small groups within engineering departments. In the two Belgian universities metrology was not an area of research but expertise in other fields was transferred to metering problems. Most university groups had fewer than ten members. Their membership included lecturing staff, research assistants and graduate students. NEL had the largest number of researchers. Flow comprises the major part of the activities of the Flow Centre measurement and is the largest single activity in the laboratory.

1.2   Research funding

The most important feature of the pattern of research funding was the small number of national research council and EU grants and the high level of industrial funding. Industry funded research in four of the seven UK departments, including both national laboratories, all three institutions in France and both universities in Belgium. Only two universities, one each in the UK and Belgium, and NEL participated in EU programmes. The balance of the different sources of income varied between the countries. French and Belgian institutions relied more heavily on industrial funding than those in the UK where the majority of academics were also undertaking research which did not involve industry. One UK respondent said that the best form of interaction with industry was that under research council funding as it allowed academics more freedom to pursue their own lines of research. Only two departments, both in the UK, had received gifts of equipment. In one case this was from Foxboro when it closed down its Redhill plant.

Departments in the UK had a wider range of sources of funding than those in France and Belgium. Two university groups were funded by the DTI to undertake applied research, both of which also had FLOMIC projects. For one group, most research funding came from the DTI. For one of these more than 50 per cent of industrial work came through FLOMIC rather than directly from industry. One of the universities, Cranfield University, was a special case in that industrial funding was the only source of research income. This was because it was then directly funded by the Department of Education and Science, unlike most other universities which were funded by the Universities Funding Council (UFC). Until 1993, it was not able to recover research overheads from SERC. This meant that it was not economically viable to undertake research council projects. This situation changed in 1994 under new EPSRC funding rules. Unlike in the electronic components industry (see Chapter 8), gifts of equipment were not a common source of income.

The French joint university/CNRS department received more funding for research from the CNRS than if it was an ordinary university department. Funding was tied to industrial collaboration. The academic interviewed explained that part of the CNRS strategy was that an increasing share of research funding and contributions to the costs of large equipment should come from industry. This meant that it had become politically important for universities to be seen to be undertaking industrial research. The consequence was that some laboratories were only undertaking applied research. Although the research activities of universities and industry had become integrated as a result of technical co-operations, this was not strictly on a commercial basis because the state was subisidizing university research funded by industry: firms applied to the Ministry of Industry for funding for the projects they placed with universities.

In Belgium, industrially funded research had increased since the 1970s as departments sought to overcome the problems of decreased levels of state funding for research. One department's specializations and special facilities had been funded either by overheads on industry grants or on EU grants such as ESPRIT. The university did not have enough money to buy expensive equipment. All staff, including tenured staff, worked on contract research. Each of the four professors in the group made their own decisions about which industry contracts to take. In spite of the high level of overheads charged on industrial projects, the department was still cheaper than private consulting firms. The main factor was that the time of the academics was not charged for. The academic said that this was a classic way in which the university system worked in Belgium: the state subsidized research undertaken for industry. Moreover, research had to be cheap to attract money from industry at all. On occasion the group would undercut competition from private testing laboratories and consultants.

All three of the national laboratories were funded to undertake industrial research, either directly or, as in the case of the UK, by the DTI. Industrial funding was the dominant form of research income at CERT, accounting for 90 per cent of total research income. Contracts came mainly from major national flow measurement instrumentation users for example from Gaz de France and from oil and gas companies such as Elf Acquitaine. There had been no contact with one of the two French gas meter manufacturers for over ten years and none at all with the other.

1.3 Geographical pattern of links with industry

In all three countries, direct links with industry were mainly inside the home country. Only three departments, two in the UK, a university and NEL's Flow Centre and one Belgian university, were funded by industry in other countries. A major factor leading to NEL's increased geographical scale of links was the decline of the Scottish industrial base. This meant that NEL had to go further afield for business. By 1992 about 10 per cent of income came from abroad. A higher percentage than this was funded by the UK branches of multi-national companies which meant that licenses were taken up by non-UK companies. The respondent commented that it was difficult to judge how much NEL's expertise was used outside the UK. Some of its contact with mainland European companies was through the BCR programme.

CERT had indirect links with German companies through Gaz de France.

1.4   Major changes

The general pattern was that in the last ten years there had been an increase in the level of contact with industry. This was associated with the need to compensate for decreased levels of state funding and political pressure to meet the needs of industry. A UK academic identified five trends:

  (i) funding problems in universities causing them to try to overcome them through collaboration with industry;

 (ii) the increased desire on the part of universities to do something useful;

(iii) the thrust from the government through various schemes such as ALVEY and LINK;

(iv) industry's tendency to cut down its own research departments and establish longer-term links with universities rather than doing it in-house;

 (v) the relative decline of government funded research labs (which the respondent felt were never really very successful at undertaking speculative research or blue sky research; as they lacked the key aspect of long-term/blue sky research which is the ability to get hold of bright young research students to do all of the thinking).

A sixth factor, was the creation of FLOMIC in 1986. This academic's department was the only one of the sample which was not a member of FLOMIC.

The main change in the UK national laboratories was the move away from academic to more applied research. NEL used to be informally known as the 'University of East Kilbride'. NEL's particular role in research in relation to industry and universities in the early 1990s was that it carried out major research programmes based on government funds, usually in consortia, some of which were carried out with universities. Sometimes this was under contract to undertake research which was further from the market. The amount of research had decreased in recent years with the change to agency status. NEL's role became more focused on the application of research in order to solve industrial problems. The respondent said that the DTI would not fund research which allowed scientists and engineers to be at the forefront of technology. The changes in the late 1980s sharply accelerated the processes begun in the

1970s by which NEL was becoming more commercial. The trigger was when it was announced in 1988 that NEL was to be commercially viable. After that point NEL had to earn money by undertaking contract research. From then it received no subsidy from the government to support anything which industry paid for directly. In 1991, 75 per cent of funding for flow measurement came directly from the DTI.

The laboratory organized its operations into four technical centres following restructuring: manufacturing systems (electronics and control and so on), testing (component testing), energy and environment (including heat transfer and wind energy), and the Flow Centre. Some £11 million was allocated by the DTI to make NEL more commercially viable. Although none of this went to flow measurement research, a £2m multi-phase flow building was opened in April 1991. This meant that NEL was well established in the world market in multiphase flow technology.

The instrumentation division within INTEC, one of the AEA's nine businesses, was at the time of the study trading externally as Harwell Instruments. Until the late 1980s, the division had expertise in flow measurement. Then, as a result of the combination of the increasing competitive environment where many TNCs were manufacturing ultrasonic meters and the restrictive terms under which it operated, the division was forced to move out of flow measurement. Harwell Instruments was uncompetitive in the meter market because the division as a whole was not allowed to either set up a production facility on site or in conjunction with a manufacturing company. The respondent believed that given a more favourable environment, Harwell would have been at the leading edge of ultrasonic meter technology, the technology used in the new UK gas meter.

Although CERT was in competition with universities for contract research, there was also a complementary relationship. Some of CERT's problems of not being able to undertake 'blue-sky' research were solved by links with the engineering schools of local universities. The students were paid little or not at all for their efforts. Sometimes students from the universities spent time preparing their theses at CERT, for months or in some cases years. This led to close ties when they went into industry. For example, the main contact at Elf Acquitaine was a student at one of the schools of aeronautics at Toulouse; the respondent had been his supervisor.

The evidence from this first section is that interaction between industry in the PSSB is increasing because of increased pressure to substitute industry for national research funding. Most funding, from industry and the state, is from within the national innovation system. The exceptions to this were the national laboratories and one UK university department.

2. Forms of interaction

2.1 Forms of assistance

The PSSB respondents were given the same list of 'forms of assistance' as those in the industry samples. The responses are included in Table 6.2. Both sets of responses matched very closely. In each case the form of assistance most often given was basic scientific knowledge. Likewise new generic technology was given as the second most common response in both samples. The universities, however, cited a much greater range of applied research and practical activities than had been specified by the firms. These included specialized testing and product development (five universities, two in the UK, two in France and one in Belgium). Three others – two in the UK and one in Belgium – provided technical reports. One UK academic saw his role as getting basic physical ideas into commercial products.

In one Belgian department, there were no a priori established criteria for accepting industry money such as scientific merit. The project undertaken for the manufacturing firm in the sample to develop software for a flow computer was leading-edge research involving the development of very sophisticated software. The equipment had to take into account all possible complex variables of flow such as speed, density, viscosity.

CERT's research was primarily directed to using the different kinds of either experimental or theoretical approaches in order to understand and improve the behaviour of flow meters in typical conditions. The amount of basic or applied research in industrial projects depended on the research contract. Work on major projects with Gaz de France and Elf Acquitaine was mainly basic research, but most contracts were for applied research. The research team had 15 years experience of study into different types of flow meters, for example orifice plate, turbine and vortex, using experimental techniques with velocity, pressure measurement and computation to understand and model behaviour of the flow meter. In this time the group had produced ten PhD theses and 40 journal articles, demonstrating that much of the work was theoretical and could be published. EXA-DEBIT's prime role was the

Table 6.2   The PSSB sample's interaction with industry

Department	Forms of interaction with industry
1. Fluid engineering and instrumentation (School of Mechanical engineering)	Basic scientific knowledge, new generic technology, assistance in product development; mainly short-term applied contracts. Current projects: four projects with industry. Long-term agreement reached with water company for three year project (worth £0.25m), development of new water meter, plus funding for research students.
2. Thermal-fluid division (School of Mechanical Engineering)	Current projects: (i) use of sonic nozzles in steam (ii) use of tracers in large duct gas flows (6 months each) plus technical reports for FLOMIC. Informal contact. (iii) two phase flow in turbine meters (£70 000).
3. Engineering science	Generic research for the purpose of getting basic physical ideas into commercial products. Current project: collaboration with user and manufacturer on enhancing accuracy of measurement of meter (3 years)
4. Fluid mechanics group (Department of Mechanical Engineering)	Occasional basic scientific knowledge, generally new generic technology, assistance in product development, solving development problems, specialized testing and problem solving, technical reports for groups of companies (FLOMIC), informal contact. Most contact is on semi-consultancy basis which involve some experimental type of work testing flow meters.
5. Electrical engineering	Basic scientific research, experimental process tomography
6. Flow Centre, NEL	Applied research. NEL provides a national calibration service, and is the centre for National Measurement Accreditation Service (NAMAS).
7. Harwell	Instrumentation, industrial technologies

France

8. Thermal dynamics group Physics	Theoretical and applied industrial research, thermo-dynamic instability. Two large projects paid for on contract basis, both have run for ten years (i) Snecma: 2 people were working on vortex problems, 4 people (1 joint Snecma/Ministry of Industry CIFRE fellowship). (ii) Schlumberger in Paris: 3 year research with a general grant which paid for a Ph.D. student's thesis, under direction of the respondent and the firm. Also contract research to the academic.

Table 6.2   Continued

Department	Forms of interaction with industry
9. CERT	Some basic scientific knowledge, mainly applied research. Basic studies of dynamic behaviours of different types of flow meters Research is mainly funded by major flow measurement instrumentation users on a contractual basis. Current project (five years): to improve the parameters of flow meters for Gaz de France and Elf Acquitaine and the Gas Research Institute. Links with Paris VI and Poitiers universities.
10. EXA-DEBIT	Calibration and testing, verification of meters, some applied in research metrology for members, improvement in aerodynamics of valves
Belgium	Two types of industrial contracts:
11. Thermal dynamics within thermal engineering faculty	(i) directly for firms (ii) funded under EC programmes. Provide four services for private companies: 1. most important in terms of state of the art research is with foreign companies such as ADF, Rhone – Poulenc, for which they develop software 2. work for foreign and Belgian companies, developing technical systems 3. testing for companies, which they sometimes do in the companies themselves, or in the laboratories 4. provide technical advice, which is limited to some kind of practical report.
12. Agricultural Sciences, Applied Sciences Faculty	A 5 year contract with small company for monitoring air flow through agricultural buildings. This constitutes their entire research effort with industry

calibration and testing, verification of meters for its member companies. It undertook some applied research but was not of major significance as a source of innovation.

3.   The interface

3.1   Initiation

Collaboration and co-operation in this sample, as in many others, was facilitated by the personal contacts of individuals. The most common form of partner selection was by personal contact, which in this industry, particularly in the UK, had been assisted by the formation of FLOMIC. NEL's Flow Liaison Club was also viewed as a useful medium. Contacts more generally were made at conferences and exhibitions and by third party recommendation. It was unusual in any of the universities for contacts to come via the institution's industrial liaison office. In Belgium, the most common way of establishing links was by personal contact but because of the fragmented industry-academic innovation system, the main barrier to interaction was finding people to make personal contact with.

3.2 Duration

The length of interaction varied between short- and long-term contracts. Five of the 12 institutions, two in the UK, two in France and one in Belgium, had long-term contracts of at least three years. In the UK, two universities and NEL were increasingly working on short-term contracts. Projects at NEL tended to be even more short-term than those in universities, some lasting as little as four or five days, others lasting six months.

Most projects undertaken in University 4 in the UK were semi-consultancy types of jobs which involved some experimental work such as testing flow meters in certain conditions. The group were prepared to do mundane jobs for industry as they maintained a dialogue with industry, raised firms' confidence in the group's ability to meet their needs and increased the general level of industry involvement. In another department some relationships were very formal, but others were more informal involving communication through a note being sent or phone calls. The degree of formality depended on how long and how well people knew each other rather than size of firm. The respondent said that if advice was easy to give, then this would be given over the phone. However, it was not unusual for a company to lose interest on being told that the problem needed investigation and would cost money.

Two of the three French institutions were undertaking long-term projects. The current project at CERT with Gaz de France and Elf Acquitaine was for five years, but contracts from industry were typically from six months to a year. These were not enough to build a new facility or new measuring device. The project involved basic research on orifice plate technology. CERT's links with the two companies had developed over 15 years. At the outset, both firms would only pay for work undertaken by research students, which was not very expensive.

For the next few years the group had a series of one year contracts paid for jointly by the two companies. These were followed in recent years by longer contracts lasting two to three years as confidence in CERT grew. The respondent said that the industrialists knew that the group in CERT could be relied upon to deliver the result while they recognized that CERT's primary activity was undertaking basic research which could provide experimental or theoretical application to answering practical questions.

3.3   Movement of personnel

Recent research has emphasized the importance of the movement of staff between industry, universities and national laboratories in creating information carrying networks and strengthening ties between them (see for example Lawton Smith 1997). In this industry recruitment either way did not occur to any great extent. Two examples were given of visiting professors being recruited from industry but only one of academics being recruited by firms. The explanation given for this was that few firms were big enough to fund research which provided an interesting environment for academics.

One of the major problems for the French university group was that it was difficult to recruit academic staff and graduate students. Salaries were low compared to what can be earned in industry. The only advantage was security. A recent solution to the problem was for contracts with industry to include an additional amount of funding to top up the salaries paid to students working on the projects.

3.4   Intellectual property rights

The allocation of IPR on industry funded research depended on the rules set by the different institutions rather than on the national framework in any of the three countries. In one UK department, if the project was of genuine research interest, the department would insist on a share of IPR. If it was a well-specified test, IPR were ceded to the firm. On industrial contracts, a legal agreement was signed defining rights to results. At CERT, when three companies worked together the rights to the commercial results were shared between them while CERT owned the principles. This was a well-established rule and IPR was not seen as an issue. However, if CERT researchers wished to publish the results of the research then permission had to be sought from the industrial partners. Generally the companies had no objection to publishing in journals with papers dealing with theoretical issues, such as fluid mechanics, but publication in applied journals on flow metering required special arrangements. Sometimes this meant that the company name appeared jointly with the researchers, and on occasion instead of the researchers.

3.5 Benefits to interaction

Benefits of interaction common to all groups from working with industry were ones previously found in other studies (Charles and Howells 1992; Faulkner and Senker 1995). These included funding research that might otherwise not be undertaken, funding for graduate students, and a source of research ideas and real life examples for use in teaching.

One of the UK departments involved in a LINK project identified three benefits accruing from the collaboration.

1 As SERC was paying for part of the project, it was pulling along in its slip-stream a whole bunch of other doctoral students. This was enhancing the group's core base of knowledge in the field and this would attract future students.
2 The user firm provided the department with good equipment.
3 The group will have been able to do a slightly longer-term job developing more advanced technology than if industry had undertaken the project. The firm would have spent less and its objectives would have been less ambitious.

In a follow-up interview other outcomes were identified. At the end of the initial period of the project the group received a large contract with the manufacturer to extend the work and received three SERC/EPSRC grants on related work, including a collaborative grant with the computer science department in the same university. Three other research students were recruited to the project. The original doctoral student graduated and became employed on the project as an RA. The work had continued without the user firm, which lost interest after the company was re-organized.

3.6 Problems in interaction

While respondents in the PSSB recognized the benefits of undertaking work on behalf of industry, there were criticisms common to all three countries.

1 The unwillingness of the flow measurement industry to fund long-term research. Academics felt that industrial contracts often adversely affected the allocation time to teaching and research.

2 Manufacturers were criticized for their inability to capitalize on technologies developed for other industries, for example fluid mechanics developed for jet engines, or to make more extensive use of existing techniques, such as computational fluid dynamics. In general the academics felt that the industry had not moved forwards in the last ten years. This meant that industry lacked information on what was on offer in universities. On the specific issue of the British Gas meter design, one university researcher maintained that there was expertise which could have been transferred from other applications into gas measurement.

3 Companies were unwilling to share information on leading-edge research. The model was one of competition rather than co-operation.

Other criticisms were specific to the UK. There, problems were to do with the nature of the work which industry was prepared to fund. The least satisfactory form of interaction for one university was the repetitive work which the group had to undertake to bring in money. The tendency for research contracts to be short term and applied meant that for the UK fluid engineering department it was difficult to achieve the ideal model of research which was to develop an idea to where it has viable applications. In practice, if an academic had a 'blue sky' idea a source of funding had to be found, and one which would be willing to fund it in increasing amounts as the idea progressed. The rarity of there being such a source of funds meant that researchers were often not able to explore interesting avenues but instead were 'goal orientated'. The consequence was that ideas were not developed to where they could be exploited. The department had to be cautious in what it proposed to industry, only picking ideas which were fairly safe. This minimized risk but had the effect of stifling creativity. A major area of conflict with industry was firms' failure to appreciate that academics could not devote 100 per cent of their time to an industrial project. Academics were unanimous in criticizing conservative positions adopted by both manufacturers and users. This took two forms. First, departments could do more industrial research. One group had not been able to get large sums of money and as a result tended to ask for small sums of money. Firms were happy to pay for semi-consultancy jobs but unwilling to subscribe £5000 on a project in partnership with other firms. Secondly, industry expected universities to provide free information. For one academic, the biggest difficulty in negotiating with industry was that most industrialists had studied at university over 20 years ago and did not realize that universities had changed since then. The university was still seen as a public service. Industrial staff were keen to pick the academic's brains for free consultancy.

The emphasis on commercialization of research in national laboratories in both the UK and France brought with it a series of problems. In the UK, industry was used to getting free advice and work from NEL and Harwell Instruments and were reluctant or unable to pay commercial rates. For NEL, this attitude combined with the effects of the recession, meant that it was facing a shortage of non-government business. The respondent felt there was a danger that there was insufficient work being undertaken on the next generation of flow meters, because neither the universities nor NEL were being funded for work in this area. It also meant that facilities at NEL, which were far greater than those in universities, were not being used for the advancement of knowledge. The respondent felt that a possible consequence was that in the long term, NEL would end up with just a test-rig. A further outcome of this trend was that the diminution of research expertise had led to tensions between NEL and some universities where it was felt that NEL was being funded to undertake research for which it did not have sufficient expertise.

Harwell Instruments found that although industry was prepared to put in effort in the form of staff time, funding for development work was not forthcoming. Very little money was made on licensing technology. This was because the laboratory was usually in a weak bargaining position. The potential licensee would point out that the cost of the development of a prototype to a product was very high. This forced down the price of the licence to the point where the returns were very low. Unless the laboratory had something that several companies wanted, it was not in a strong bargaining position and it was usually the case that only one company was interested. The respondent felt that 'UK instrumentation companies are a parsimonious lot'.

Some of CERT's problems were related to how it and the universities were funded. For example, CERT was invited to participate in an EC BCR-funded collaboration with a firm and a research group from a university in Portugal. The project was to compare some velocity measurement with basic calibration in pipes under different conditions. As most European programmes only fund 50 per cent of the grant, CERT needed to find most of the other 50 per cent from industry because of the small proportion of funding CERT received from the state. The problem of finding industrial partners made it hard for CERT to enter such collaborations. Moreover, because universities were increasingly under pressure to find external funding as the level of state support decreased, CERT was facing direct competition from this quarter. Sometimes research in universities would be only a quarter or fifth of the cost for the same kind of work, because staff salaries were not included in the cost of the project to industry whereas at CERT salaries of senior academics were paid from contract research. Three-quarters of the budget of the department went on salaries. The respondent, the head of department, said that he spent 'too much time looking for money'.

4.   FLOMIC, consortia and committees

4.1   FLOMIC

As in the industry sample, most of the respondents in the UK belonged to FLOMIC. The respondent from CERT was a member for the same reasons as the engineer from Gaz de France – keeping up with technical priorities and research agenda in industry and the PSSB.

Views on the value of the organization were mixed. Some academics felt that it should have been more successful. Most were critical of the amount of work required on projects for which the budget was £15,000. The literature survey for some projects could take weeks. However, although the financial reward was not worthwhile one academic felt that the resulting contact with industry was important.

4.2   Clubs

NEL was involved in a number of joint industry/research activities. These included membership of consortia established by NEL to tackle specific problems including EUROMET and innovation clubs. NEL runs a number of innovation clubs. The most important is the Flow Liaison Club. Its members include universities, and other higher education institutes, manufacturers and users of instrumentation. Meetings are hosted in different places. Activities include round table discussions and presentations by group members or by people from outside. It does not commission research but acts as a 'talking shop'. In 1991 it had some 80 members, of which 20 were companies. Attendance at meetings averaged about 50. NEL also runs a multi-phase flow consortium, dealing with all topics related to multi-phase flow. This is entirely funded by the North Sea industry.

4.3   Committee membership

A pattern common to both PSSB and industry samples was the limited involvement on committees which decided the allocation of research funding. Respondents from UK institutions were more likely to serve on standards than on research council committees. CERT had access to the policy making process because one of the department's professors was on the research committee of the Ministry of Industry.

There were some national differences in the membership of standards committees. Two of the UK institutions, a NEL and an academic were members of the BSI flow Measurement Committee. In contrast in France academics tended not to sit on standards committees. These were dominated by companies such as Gaz de France and Elf Acquitaine.

5. Conclusions

The main finding was that the general increase in interaction between academics, laboratory scientists and engineers in all three countries was a direct response to the changes in the regulatory environment. Pressure to raise the amount of work undertaken for industry had been brought by reductions in state funding for university research and by the institution of a commercial rather than a research ethos to the activities of national laboratories.

In most cases interactions can be seen as examples of externalization rather than collaboration. This is because academics were engaged in short-term problem solving activities using existing expertise rather than undertaking studies which required significant technological advances. There were a few examples of longer-term collaborations with industry, for example at CERT and in the UK under the LINK programme. Most interaction was within the home country. Few departments were involved in international collaborations either on a one-to-one basis with industry or under EU programmes. FLOMIC could have provided a medium for networks to develop over national boundaries but at the time participation was limited to France and the UK.

While research staff in both kinds of institutions had no choice but to accept the reality that research and industrial funding were linked there was still the problem expressed in the old adage 'you can take a horse to water but you cannot make it drink'. It appeared that the problem was not that there was insufficient expertise in the science base but that industry was unable or unwilling to capitalize on opportunities available. Academics in all three countries were frustrated by the failure of industry to look for and identify the commercial possibilities offered by the technical resources of PSSB. This finding echoes those in other studies, such as Charles and Howells (1992) who also found reluctance on the part of industry to pay for research. Moreover, even when contact had been established, firms were unwilling to pay what academics thought was a realistic amount. They were frustrated by the dominance of financial rather than technological considerations when projects were proposed. The academic ethos of making information freely available was therefore in conflict with the need to commodify technology. FLOMIC had been unable to bridge the gaps between the needs of the academic community on the one hand and industry on the other. In addition, the lack of mobility between the PSSB and industry meant that a standard mechanism for establishing and re-inforcing links, and strengthening filieres, was largely absent. A solution to the problem was suggested by one UK academic. He thought that the attractiveness of interaction might be improved if universities could offer a portfolio of activities which included short-term research combined with a longer support for a student or research fellow. This would have had the effect of raising the platform of innovation in industry, enabling it to interact more with universities.

The consequence of NEL moving to more applied activities and its increasingly dominant role of providing testing services was that its central position within the national and international system of innovation in metrology was declining. For Harwell Instruments this had happened some years earlier. On the other hand although CERT was the most dependent on industry for funding, it was able to maintain a research profile which included theoretical as well as more applied activities.

Both NEL and CERT, but not the universities, were becoming more international in their activities. In the case of NEL this was through formal international institutional arrangements for which it was the UK representative, and through consortia.

The most important regulator of interaction therefore was that the balance of power lay with industry. Firms were able to dictate ownership of intellectual property and 'free-ride' the state system. The reality was that academics had very little bargaining power with industry. Not only was there a need to bring in industrial funding in order for research of any kind to take place, but individuals were affected by the politicization of the innovation filiere. Promotion chances would be improved with a large contract on the CV. The effect of this pressure was that academics were more likely to accept what came their way and waive rights to intellectual property. The consequence was that research which was theoretically significant but which had no obvious short-term payback was ignored. This meant that creatively was stifled and that there was an under-utilization of research skills and equipment. Belgian academics were even more dependent on industry than those in the UK and France. There the state system underpinned industrial interaction by undercharging on commercial criteria for the use of resources.

7
Externalization Patterns in the Electronic Components Industry

Introduction

This chapter examines linkages between firms in the electronic components industry and universities and national laboratories. It follows the same structure as Chapter 5. It highlights the main factors which regulate the characteristics and strength of the filieres in this sector in each country. It also indicates some of the main differences between this sample of electronic components firms and that of the flow measurement industry.

1. The sample

The electronic components sample consisted of 23 firms: 11 in the UK, five in France and seven in Belgium. Details of the firms' ownership, size, date of formation, ownership, activity and markets are given in Table 7.1. The sample included small independent firms, domestic and foreign-owned MNCs. Most firms in the UK and France were domestically owned. In contrast, ownership in Belgium mirrored the national pattern of foreign investment, only one firm was Belgian, the rest had headquarters elsewhere – in France, the Netherlands, the US and Germany. As in the flow measurement industry, there had been considerable take-over activity in each country, only a very small number of sites had been established by foreign MNCs. The sites visited ranged in size from 26 (an independent UK firm) to 1600 (a manufacturing plant in Belgium). Most had been formed post-war.

The sample consisted of manufacturers and research centres. Manufacturers comprised three groups: those that did not all

Table 7.1   The electronic components industry sample, employees, date of formation and major markets

Ownership	Employees	Date formed	Main markets
UK			
1. UK	26	1983	Semiconductor mfrs, UK 43 per cent, rest of Europe 40 per cent, electronic companies and universities. Competitors – US
2. Japan	32 (65,000)	1990	Consumer electronics, business electronics and consumer electrical equipment. Japan and USA, Europe (less than 10 per cent is within the UK). 130 customers are all 'blue-chip' electronic device manufacturers. Competitors MNCs and small UK manufacturers (research centre)
3. UK	88	1988	Semicon. crystals, Japan, Taiwan
4. UK	4000 (70/80,000)	1980	Semiconductors, space industry. Markets are the US (40 per cent), UK (40 per cent) and the rest of Europe (20 per cent), competitors – US and Japanese
5. Holland	500 (1500)	1975	All kinds of circuits, consumer, industrial, automotive, lighting industries. Some 40 per cent is sold in the rest of Europe, 35 per cent in the Far East and North America 12 per cent, competitors in Japan, France, US
6. USA	160 (2000)	1979	Electronic connectors and electrical fittings, distributes via sister companies, major competitor – UK firm
7. UK	115 (1300)	1981	Micro-electronics companies, universities and research institutes UK market 10 per cent, Europe and the US 30 per cent each, and Japan and the Far East together 30 per cent. Competitors – USA
8. Canada	900 (32,000)	1960	Telecoms, competitors – MNCs (research centre)
9. Italy/ France	700 (MNC)	1978	Microprocessors, competitors – major microprocessor manufacturers.

Table 7.1   Continued

Ownership	Employees	Date formed	Main markets
10. USA	22 people in design centre (MNC)	1969	DRAMs microprocessors for telecoms, automobile industries, competitors in Japan
11. UK	215 (MNC)	not known	Military, international competitors (research centre)
France			
12. France	40 (55)	1927	Transformers, power industry, transport, audio-visual industries, medical, Turnover FF 12 million 1991
13. France	not known (MNC)	not known	Very wide range of electronic components for space and telecoms, world-wide competition (research centre)
14. France	100 (800)	1946	Delay lines. World-wide some 20/30 competitors including USA, Japan and the UK, no domestic competitors.
15. France	400	1988	Printing technologies, competitors, Japan
16. USA	1300	1938	Resistors, consumer electronics, telecoms, computers, defence etc.
Belgium			
17. France	120 (7000)	not known	Telecoms, competitors world-wide (research laboratory)
18. France	500 (7000)	not known	Telecoms. Half of output is transferred to a sister plant in Wallonia, another 30 per cent is delivered to sister companies in mainly in Europe and the remaining 20 per cent is sold in the open market for military, telecoms and industrial applications in Western Europe. Less than 5 per cent in sold in Belgium on open market. Competitors sister houses in Germany and the Middle and Far East

Table 7.1   Continued

Ownership	Employees	Date formed	Main markets
19. France	50 (450)	1983	Telecoms. In 1991, total sales of $105m Some 70 per cent within the group and the rest to the open market. The intention was to decrease share to 60 or 50 per cent by increasing the merchant market. Rest – mainly other telecoms, automotive and industrial companies, some specific applications in consumer industries, competitors world-wide
20. Netherlands	150 (MNC)	not known	Motor industry e.g. small stepper motors, circuit breakers, sensor, also packaging for hybrid circuits. Uses: automotive and industrial applications – mostly infomatics – computers. Turnover 500 million BF 1991, about 90 per cent of sales are in Europe. Japanese and European competitors, mainly German, no UK or French
21. Germany	1600 (MNC)	not known	Telecoms (internal use) Typically 25–30 per cent of sales in export business. Competitors – many small local companies
22. Belgium	2500	c1962	Displays (internal use) industrial and consumer use, competitors – local plus Netherlands and France Turnover 10 million BF 1991
23. USA	300 (MNC)	1960	Car industry 600 million BF in 1991. Export almost 90 per cent of main products major markets in France, Germany and Holland. Small markets in the UK, Italy and Spain. Competitors in hybrids and thick film mainly USA, decreasing number of large and small UK firms in hybrids. Internal supply from big firms.

undertake R&D (all of which were branch plants of foreign-owned companies); leading-edge manufacturers of state-of-the-art components; and manufacturers of specialist equipment and materials. In the UK sample there were three research centres of major international companies, and one each in France, the central laboratory of one of the largest French electronics companies, and Belgium the research facility of a formerly Belgian, now French, telecoms company. In the UK the centres were those of a UK aerospace company, a Canadian electronics company and the European research centre of a Japanese electronics firm.

Product areas included semiconductors, ASICs, PCBs, hybrid circuits, electronic connectors and materials. Markets included domestic electronics, telecoms, space, automotive industry, transport, medical applications, business electronics, university research and industrial applications. The majority of the firms competed in international markets: only three firms, two Belgian and one UK, mentioned competition from local firms.

2. Innovation strategies

There were considerable variations in R&D activities within the sites visited. They ranged from long-term research activities in research centres to engineering-based product development in branch plants. The general difference in innovation strategies in this industry to that of the flow measurement industry was that whereas none of the flow measurement sample set out to out-compete other firms by the introduction of radical advances in technology, technological lead was the basis of competition for most of the firms in the sample. However, even in the most technologically dynamic manufacturing firms, there was a clear pattern that price competition and the necessity of rationalising production into few product areas were driving innovation strategies.

3. Interaction with universities and national laboratories

3.1 Extent of linkages

The central hypothesis of the study was that there would be a general increased interaction with universities and national laboratories. This hypothesis was supported up to a point by evidence from this sample as several were becoming more active in seeking sources of technical support from the PSSB. On the other hand, as in the flow measurement sample, some firms were reducing their links with the PSSB. The overall level of interaction was high. The majority of firms (18; 78 per cent) had links with at least one university, although rather fewer had research relationships with national laboratories (9; 39 per cent) (Table 7.2). Most had informal contact even if there were no formal links. There were some similarities and some differences between the countries. Generally the UK and Belgium samples showed similar patterns of interaction and level of public subsidies, whereas the French sample had the fewest formal links and the lowest level of both national/regional and EU grants.

A pattern common to all three countries was that firms without formal links were branch plants of MNCs. However, even within this group there was a general recognition that universities might be important in the future. Only one in France had not considered working with universities.

The main difference between the UK and the two other countries was that firms in the UK were more likely to have links with national laboratories than those in France and Belgium. This is likely to be a reflection of the composition of the samples in France and Belgium because interviews with LETI and IMEC suggested that these laboratories had extensive links with firms in their respective countries.

Universities

In recent years there had been a number of changes in strategy towards involvement with universities. Firms are categorized as follows:

1. Those which were re-focusing their links   These were firms which had always worked with universities (five in the UK, one in France and two in Belgium) but which had begun to strengthen existing relationships with a reduced number of institutions, to spend less on speculative research and more on specific problem solving and to place more emphasis on recruitment. For three UK owned firms this meant an overall decrease in the level of interaction.

2. Firms which were decreasing levels of interaction   This includes:

- firms in the UK which had withdrawn from semiconductor manufacture
- A Belgian firm facing an adverse competitive environment was cutting back on R&D spending.

3. Firms which had recently begun to look to work with universities and to recruit graduates   These were:

- an independent French firm in France was motivated by the demands from customers to produce new technical solutions. Improved profitability had meant greater resources for R&D
- a US semiconductor manufacturer located in Scotland's R&D centre was in Texas. Its links with UK universities were limited but increasing as it had started to look to recruit graduates and had only recently begun to consider the possibilities for funding research.

4. Firms which were looking to increase their use of university research. This group included three firms in the UK with corporate headquarters overseas (in Japan, Netherlands and Canada).

The kinds of changes instituted by the first group are illustrated by three examples.

Example 1   For one UK firm, R&D priorities had been rethought following the merger with another UK electronics firm. Under the new arrangements, the range of technologies and product lines had been reduced. The strategy adopted was to concentrate on a small market area with competition was limited to a small number of firms where the company had special expertise, design and manufacturing capability and had made the right level of investment to maintain its position. The decision to go for niche markets was at the expense of maintaining a position in the bulk semiconductor market. The firm had withdrawn from a major programme on the next generation of semiconductors largely because of the level of investment required. Merger had brought closer integration of R&D and manufacturing and some of the research from the central laboratory (since closed) had moved to production sites. A second outcome of the merger was that the additional high quality resources available in-house meant that instead of looking for partners outside the company, research collaboration was more internally focused. Some programmes of research had changed as a result which in turn meant less interaction with universities.

Example 2   An important change for one of the manufacturers in the French sample over the previous five years had been that as a result of technological advances in the next stage in the production chain, its customers required smaller and fewer transformers. This required changes in design and production of components. The R&D centre identified that technology was increasingly a key to success, combined with price. The trend was to make components faster and smaller and to use R&D in a global marketing approach. The respondent said that 30 years ago this was a technologically driven company where scientific people had freedom to innovate. There had been a gradual change and the company had become much more market orientated. Technology had become more international, competition was fiercer and there were fewer protected markets, such as defence.

The R&D efforts of a Belgian manufacturing firm were directed to improving production processes as a means of increasing profitability. This part of the firm's innovation activities did not require input from universities or national laboratories.

Example 3 Until the early 1990s the Belgian research centre of a French telecoms group links with universities (and IMEC) used to be open-ended and speculative but had changed to being directed to specific problem solving. Research activities used to be based on good relationships with individual academics. More recently, selection was based on which was the best university in a particular domain. Recruitment from universities and IMEC remained very important. The importance of cost as a criterion for using universities as an extension of in-house research is provided by this example. The respondent estimated that university researchers cost about 60 per cent of company

Table 7.2 Electronic components sample's links with universities and national laboratories and government funding

	Sample size	University links number (per cent)[1]	National laboratory link number (per cent)	National/ regional support number (per cent)	EU support number (per cent)
UK	11	9	6	8	6
		(82)	(55)	(73)	(55)
France	5	3	1	2	1
		(60)	(20)	(40)	(20)
Belgium	7	6	2	5	3
		(86)	(29)	(71)	(43)
Total	23	18 (78)[2]	9 (39)	15 (65)	10 (43)

[1] percentage of national sample
[2] percentage of total sample

manpower. However, the centre thought it was losing the other 40 per cent because the firm had no direct control over the technology and universities were slow in meeting deadlines. If costs went up, it would consider further decreasing its work with universities.

National laboratories

National laboratories were important in the innovation strategies of the most R&D intensive firms in each country. In the UK about half, some of which were SMEs, undertook joint research with laboratories including Harwell, RSRE Malvern, LETI and IMEC. Links ranged from licensing, contract research, collaborations and recruitment. The UK Japanese research centre had recruited two people from what was then RSRE Malvern, both of whom maintained their links informally.

In France, the electronics research centre had links with several national laboratories. The advantage of working with LETI derived from it being much more technology driven than both French universities and the CNRS. Another firm used national laboratories for testing or standards development, such as the Laboratoire Centrale Des Industry Electriques (LCEO). Two sites of a French telecoms group in Belgium had links with IMEC.

3.2   Geographical scale of linkage

The innovation filieres in each of the three countries in this sector were far more spatially extensive than those in the flow measurement industry. Firms were more likely to have links with several institutions in the home country and with ones in other countries. Half had some form of interaction with foreign institutions. The pattern was not related to size or usually, to ownership. Small firms were as likely as medium and large firms to have links with several institutions. Those firms which were producing the less-high tech components such as connectors and delay lines either had no links or had low-level links. The status of the plant was important in the case of two US branch plants in the UK. In both cases R&D was conducted in the US. One was the semiconductor manufacturer in Scotland, mentioned above, where R&D was conducted in the US, the other was a manufacturer of connectors and electrical fittings.

There was a difference between the three countries in the number of institutions named by firms. The UK sample (both manufacturers and research centres) tended to have far more links than firms in France and Belgium. Most had contacts with at least ten institutions. The sample named 34 UK universities. Oxford University was named most often, followed by Southampton and Edinburgh. The norm in France and Belgium, countries with fewer centres of expertise, was for firms to have links with less than five.

There were similarities in the extent to which manufacturing firms in the UK and Belgium had links with foreign institutions. Half of the UK and Belgian manufacturers had links with foreign firms. In the Belgian sample all of these were in the telecoms sector. This is associated with the composition of the sample which reflects the local specialization in Flanders in telecoms. None of the French manufacturers had links outside France. The research centres in all three countries displayed similar patterns in that they all had links with both domestic and foreign universities and national laboratories.

More evidence of localized interactions was found in France than in the UK, although firms in the UK usually, but not always, had links with 'local universities', those within the same region or adjacent region, as well as with institutions elsewhere. Local links were being developed in two cases. These were the subsidiaries of foreign firms which were in the early stages of interacting with universities (Firms 5 and 10). In France all had links with institutions within their own region, only two had additional contacts in other parts of France. The majority of the Belgian sample had links with the University of Ghent (Flanders) but firms tended to have contact with all of the six major universities in both Flanders and Wallonia.

The pattern of non-national links was that they were mostly within Europe. Four UK firms and the French research centre collaborated with universities in North America. The Japanese research laboratory based in the UK had begun linking up with German and French universities as part of its recruitment strategy and plans to gain access to European technology. One UK manufacturer which was looking to reduce the number of UK institutions with which it worked while increasing the range of universities and national laboratories on mainland Europe. One UK firm had long-established contacts with LETI but none with UK laboratories. It had found LETI to be very industry orientated and had excellent facilities, unlike UK institutions. The firm benefited from the kudos of its links, especially in France and from keeping abreast of what the 'big players' were doing. Another had a formal research arrangement with IMEC.

3.3 Motives

Motivations of firms recorded for initiating and sustaining externalization of R&D activities reflected both the technological trajectories of

Table 7.3   The electronic components industry sample's links with universities and national laboratories, and their location.

Ownership	University links	Notes
UK		
1. UK (North West)	UK: Oxford, QMW, Liverpool, St Andrews, Glasgow, Surrey, Reading, Strathclyde, Harwell, RSRE Non-UK: Berlin, Munich, Montpellier, Regenburg, CNRS	Long term, UK funding and EU initiatives
2. Japan (South East)	Cardiff, Exeter, UCL, Oxford, Imperial, Sussex, King's London*, Germany, France	CASE students at 4 universities
3. UK (South West)	Formal UK links: Cardiff, Glasgow, Loughborough, Bristol, King's London, Sheffield, Cambridge, Southampton, Durham, Nottingham, Imperial, UMIST, Surrey, Oxford. Formal non-UK Links: Nijmegen, Africaans South Africa. Informal UK links: Oxford, Bath, Hull, Leeds, Ulster Informal non-UK links: Michigan, Alabama, BC, Cincinnati, California, Texas, Montreal, Illinois, Ghent, Denmark	Long term
4. UK (East Midlands)	UK links: Sheffield, Southampton, Middlesex, Bradford, Edinburgh, RSRE, USA	Long-term collaborations mostly under UK and EU initiatives
5. Holland (North West)	UMIST Salford	Six month project
6. US (South East)	Nil	
7. UK (South West)	UK: Cardiff, Edinburgh, Salford, Leeds, Glasgow, Southampton, RSRE. Non-UK: LETI, Kassel, MIT	Long-term collaborations usually under UK and EU schemes
8. Canada (South East)	UK: Southampton, Reading, Oxford, Cambridge, Imperial*, UCL, Nottingham, Sheffield, Edinburgh, Glasgow, Manchester, Liverpool, Swansea, Bristol, Surrey, RSRE Malvern	Long-term collaborations under LINK and ESPRIT projects
9. Italy/France (South West)	UK: Oxford, Cambridge, Manchester, Southampton, Liverpool, DRA Non-UK: Karlsruhe, TA Munich, UPM Madrid, CTI Patras, SICS, INESC	Long-term collaborations usually under UK and EU programmes

Table 7.3 Continued

Ownership	University links	Notes
10. US (Scotland)	Edinburgh	Three year links through CASE
11. UK (South West)	UK: Oxford, Edinburgh, Bristol, Bath, Imperial*, Cardiff, Swansea, York, Newcastle, RSRE Malvern	long term

France

12. France (Paris)	Angers	Based on training link
13. France (Paris)	French Institutions; CNRS, CEA, INRA, LETI Universities in Paris and Grenoble/ Lyon regions Non-French Institutions: Cambridge, MIT	Formal arrangements with overseas universities
14. France (Paris)	nil	
15. France (Paris)	Clermont Ferrand, Brittany, Limoges, Paris	Technical discussions, recruitment of engineers, attend seminars as a means of making contact, testing in national laboratories
16. USA (Provence)	Nice/Grenoble areas	Look mainly for local support, but would use other universities in Paris and the north of France

Belgium

17. France (Flanders)	All major universities in Belgium, IMEC, European universities	Long term, EU programmes e.g. ESPRIT, Flanders support
18. France (Flanders)	Informal: KUL, Ghent Berlin	Research links are based at main research centre in Antwerp
19. France (Flanders)	Ghent, KUL, IMEC, Louvain-le-Neuve	Long term, externalization of R&D to IMEC, (4 year contract) long-term interaction with Belgian universities, ESPRIT

Table 7.3   Continued

Ownership	University links	Notes
20. Netherlands (Flanders)	nil	Links with universities based at main research lab. in Netherlands
21. Germany (Flanders)	Ghent, Liege, Mons	Special relationship with Mons
22. Belgium (Flanders)	None current	In past had projects with KUL, Namur, Brussels, Ghent
23. USA (Flanders)	Ghent, Limburg, Antwerp	Graduate scholarship received information about presentations made at IMEC.

*University of London colleges are included separately

individual firms and the kinds of contextual factors identified in Chapters 2, 3 and 4. These include such factors such as the key role of particular research institutions within the national system of innovation such as the Malvern laboratory, the availability of public subsidies and the research intensity of individual sites of firms. The two most common motives for collaboration with universities in all three countries were related to the necessity of gaining access to a new area of expertise and to existing information. This pattern was repeated for national laboratories although with a much lower frequency. Whereas the most common response in the flow measurement sample was inadequate scale of R&D, this was much less of a factor in this industry.

The main difference between the countries was the higher importance accorded to financial considerations in the UK and Belgium than in France. More than half of the Belgian and a third of the UK samples said that cost saving was an incentive to collaborate with universities. This was important in whether firms would work with national laboratories in only the UK. This is a very similar finding to that in the flow measurement sample. The finding that over half of the UK sample cited participation in public programmes as being an important motive for working with universities and a quarter with national laboratories is consistent with the general pattern that the UK has the largest number of EC RTD collaborative links of any member state (Georghiou *et al.* 1992, i).

Table 7.4 The electronic components sample's motives for collaboration with universities and national laboratories

Motives for collaboration	UK	France	Belgium	Total
Universities				
Cost saving	4	0	3	7
Inadequate scale of internal R&D	3	2	2	7
Standards development	0	0	0	0
Access to new area of expertise	5	4	3	12
Access to existing information	2	4	4	10
Strategic benefit/opportunity	3	0	0	3
Participation in public programmes	6	2	1	9
Access to public sector markets	0	0	0	0
Complementary expertise	3	0	1	4
Internal problem which could not be solved	4	0	1	5
Other	2	0	1	3
None	2	1	3	6
National laboratories				
Cost saving	2	0	0	2
Inadequate scale of internal R&D	3	0	1	4
Standards development	1	1	0	2
Access to new area of expertise	3	1	1	5
Access to existing information	1	1	2	4
Strategic benefit/opportunity	1	0	0	1
Participation in public programmes	3	1	0	4
Access to public sector markets	0	0	0	0
Internal problem which could not be solved	1	0	1	2
Other	3	0	0	3
None	4	3	4	11
Total sample size	**11**	**5**	**7**	**23**

3.4 Forms of assistance

The main function of universities for this sample as a whole was to provide basic scientific knowledge and to assist in the development of new generic technology. These were the priorities in France and the UK. Universities and, to a lesser extent, national laboratories frequently made contributions to new product development and solving product development problems. UK universities but not those in France and Belgium helped with routine testing or validation services. Table 7.5 shows the responses of firms to the question about the forms of assistance they have obtained from both universities and national

Table 7.5   The electronic components sample's forms of assistance with universities and national laboratories

Forms of assistance	UK	France	Belgium	Total
Universities				
Basic scientific knowledge	8	4	2	14
New generic technology development	6	3	3	12
Assistance in new product development	3	3	3	9
Developing testing routines	3	2	1	6
Solving product development problems	4	2	2	8
Improving production processes	3	1	0	4
Routine testing or validation services	3	0	0	3
Adaptation of technology to local conditions	0	0	0	0
Other	1	1	1	3
None sought	1	1	3	5
National laboratories				
Basic scientific knowledge	1	1	2	4
New generic technology development	2	1	2	5
Assistance in new product development	3	1	1	5
Developing testing routines	1	0	1	2
Solving product development problems	1	1	1	3
Improving production processes	0	1	0	1
Routine testing or validation services	1	1	0	2
Adaptation of technology to local conditions	0	0	0	0
Other	2	0	0	2
None sought	3	3	5	11
Total sample size	**11**	**5**	**7**	**23**

laboratories. Conditional factors which regulated the forms of assistance given included whether universities had the right equipment and whether they understood the company's needs.

The pattern in France was similar to that in the UK. Universities provided basic scientific knowledge for four companies and new generic technology for three. Universities in France also worked on short-term applied projects. Three firms obtained help with new product development, one of which was assisted with solving product development problems. In Belgium, the focus was more on applied assistance rather than on acquiring basic scientific knowledge. Firms gave a much more limited range of responses to those in the UK and France, suggesting that relationships in Belgium were more focused than in the other countries.

3.5 Duration

The duration of links varied between the countries. They tended to be longer in the UK than in France and Belgium. In France, two of the three firms which interacted with the French higher education sector had only short-term links of no more than one year. In Belgium, three firms had short-term or occasional links with universities. Longer projects were confined to two members of a French telecoms company. Part of the explanation of the differences is the pattern of government/EU subsidies. Joint projects organized under formal programmes were usually longer than those funded directly by industry.

3.6 Funding

A characteristic of the electronic components industry was the high level of national/regional/EU support for research interaction (see Tables 7.2 and 7.3). There were, however, some national differences (Table 7.6). Nearly three-quarters of UK (73 per cent) and Belgian firms (71 per cent) had obtained grants under national/regional programmes. In France the percentage was much smaller (40 per cent). In spite of the importance of defence in the UK and French national innovation systems, MoD funding was not a major source of subsidy. Only one UK firm, and two in France, had links with the PSSB supported by the MoD.

Further differences were apparent in EU funding. The UK had the highest percentage of firms with EU awards, France had the least. Part of the explanation is the differences in the composition of the French and UK samples. There were proportionally more 'high-tech' firms in the UK which were more likely to participate in international collaborations and more 'low-tech' firms in France which were not. Size was not associated with whether firms were funded by either UK or EU programmes. Both large and small firms participated in UK programmes such as CASE (seven firms, 64 per cent) and LINK (six firms, 55 per cent). Half were in ESPRIT collaborations. Smaller firms were frequently invited into programmes as specialist suppliers. Only two firms, branch plants with headquarters in the Netherlands and the US, had no public funding. The two Belgian firms with EU awards belonged to the same French telecoms firm.

UK firms with LINK projects were highly critical of the scheme, as they were in the UK flow measurement industry. The 50 per cent funding rule was seen as having the effect that the larger the university share of project expenditure, the smaller the net grant to industrial partners. Several respondents made the same point: that in this

Table 7.6   The electronic components industry sample research support

Ownership	Employees	National support	EU support
UK			
1. UK	26	4 CASE awards with different universities, 6 LINK, 2 DVCD	1 BRITE/EURAM, ESPRIT
2. Japan	32	CASE	
3. UK	88	Several LINK, GaAs HEMPT Programme, CASE	SPRINT, ESCOR, ESPRIT, ESA
4. UK	4000	MoD	ESPRIT
5. Holland	500	nil	nil
6. USA	2000	CASE	nil
7. UK	115	5 CASE students, Teaching Company Scheme	ESPRIT
8. Canada	MNC	LINK	ESPRIT
9. Italy/France	700	LINK, CASE	ESPRIT
10. USA	160	nil	nil
11. UK	215	LINK, CASE, MoD	EC
France			
1. France	40	nil, unsuccessful application for M. of Ind. funds	nil
2. France	MNC	MRT, MoD	ESPRIT, BRITE/EURAM, RACE, EUREKA
3. France	800	Unsuccessful applications to Ministry of Industry and ANVAR, MoD	nil
4. France	400	MRT, Industry Ministry, ANVAR	EUREKA
5. France	1300	nil	nil
Belgium			
1. France	120	IWT, Telecoms Fund	ESPRIT
2. France	500	IWT	JESSI, ESPRIT
3. France	450	IWOLN	nil
4. Netherlands	150	nil	nil
5. Germany	1600	IWT, Flanders Telecoms Fund	nil
6. Belgium	250	nil but have had in the past	nil
7. US	300	Flanders Telecoms Fund	nil

situation there was a danger was that universities were brought in just to satisfy LINK conditions and not used effectively.

In France, smaller firms had not found it easy to gain access to public funds. Only one had been successful. Two others had failed to get funds from the Ministry of Industry. The larger firms were much more success-ful. The central research laboratory received up to 50 per cent of its R&D budget from a combination of French and EU funds. It automatically applied for funds under national rather than regional programmes such as those organized by the MRT. This pattern reflects the disproportionate share of government industrial subsidies obtained by large firms. One small firm which had succeeded with getting national funds was also looking at the EUREKA programme, suggesting that there is some spring-board or learning effect from obtaining one source of funding derived from knowing what kinds of input from firms these programmes require.

About half of the Belgian sample had funding from public author-ities for links with the PSSB. Three of the seven firms had support from the IWT and two of these and another had funds from the Flanders telecoms fund. The level of funding from public funds could be high. The research centre received between 30–40 per cent of its annual R&D funding from regional, national and EU initiatives. Another firm within the same group had recently been involved in five European programmes including ESPRIT and JESSI. In the last round of applica-tions, the firm had been named in 13 proposals although ten were unsuccessful. Several projects had involved IMEC which had played a co-ordinating role in putting together a critical mass of university expertise and acting as a single interface, thus overcoming the manage-ment overhead problem.

The following five case studies illustrate the general priorities and prob-lems associated with managing the interface with the PSSB. Three are from the UK and one each from France and Belgium. Particular points of interest are the perceptions about the relative positions of firms and uni-versities in the innovation process, the range of interactions, the use of power exerted by industry on the science base, the issues arising from interaction and therefore the kinds of interdependencies and countervail-ing forces which comprise the processes shaping the innovation filieres.

3.6 Five case studies

A. The European research centre of a Japanese electronics company

The research centre had been set up in the UK because the firm believed that it would have access to innovative people in UK universities first then to academics on mainland Europe. The firm's largest manufacturing facilities in Europe were in the UK. The focus on using external expertise to create basic technological advances in-house was reflected in the forms of assistance sought from the PSSB. It was seeking complementary expertise at a basic/theoretical level. The strategy to maximize the possibilities for information flows into the firm involved a raft of formal and informal contacts from informal net-works, secondment of staff to universities, contract research, consul-tancy, sponsoring CASE students and in the future, subsidized collaborations. It was planning to enter a DTI software project at a nearby university. The policy on IPR on contract research at universi-ties was that it wanted exclusive rights.

B. The UK research centre of a Canadian electronics firm

The centre had a range of motives for research collaboration with uni-versities, relating to quality of work, proven capability and facilities, for example ion-implantation. Linking to universities brought additions to the company's research portfolio. Three factors influenced interaction (a) personal knowledge of contacts, (b) participation in collaborative programmes, and (c) recruitment. The company rarely took up licences as it was unusual for universities to have the ability to produce any-thing that could be converted into a product. The firm did not want a formal relationship with a university on a one-to-one basis and pre-ferred contact through collaborative programmes. On the other hand it occasionally paid for small pieces of work in universities or for support for facilities in exchange for analysis of samples. The respondent believed that the company was well up-to-date on developments in both industry and higher education because it was well coupled into collaborative programmes and because senior staff served on DTI, SERC and other important advisory bodies. Interaction was important to competitiveness and market position, and had reduced the costs of adopting technology.

C. The UK aerospace research centre

The centre links with virtually every UK university but named seven with which it had especially strong contact. Interaction was through direct contract research and through formal arrangements such as LINK and CASE studentships. CASE studentships were important because they provided useful research, possible future employees and the basis for a continuing relationship with academics. The centre was undertaking some research in conjunction with universities into materials used in semiconductor manufacture, active materials for use in integrated circuits, on GaAs and micro-mechanical devices. It was also working on parallel processing with two universities, and was supporting the design of a new network chip at another. While the respondent believed universities could provide some practical applications, he did not believe that this was their correct function.

D. The French central electronics research centre

The respondent was the liaison officer in charge of managing external research linkages. For the firm as a whole, external R&D linkages were becoming more important because the real high level work could not be undertaken internally. The firm had links with most universities and national laboratories in France and collaborated extensively throughout Europe and with US universities such as MIT. All divisions of the company had contacts with universities, creating very complicated networks of collaborations. Contacts were directly with individual departments and through EU programmes. Close links had developed with the Ecole Polytechnique which was 200 metres from the laboratory. University staff worked on R&D in the firm's research laboratory and some of the firm's senior researchers lectured in the school. EU collaborations were very important and would be more so in the future.

Although contacts were extensive the amount of money the centre was putting into CNRS/universities directly was very small. The more usual way to solve the financial problem was to jointly apply for support from French administration such as the Ministry of Research and Technology and the Ministry of Defence. The firm had a general framework of collaboration with CNRS, CEN and INRIA laboratories. A general agreement covered industrial patents. For special projects, all the laboratories in the divisions would fund dedicated programmes. It was unusual for a division to ask a laboratory to solve a specific problem; where it did it would pay the institution directly. More frequently there was a long-term relationship with a CNRS, INRIA or CEA laboratory.

The firm identified a number of problems it faced in working with the PSSB. The respondent said that CNRS laboratories were very good at fundamental research such as mathematics and physics and biology, but not so good at technology as there was not the tradition of technology laboratories as in the US. In France it was a question of culture. Good students wanted to be a theoretician first, failing that they wanted to be an experimental but not an applied physicist. This was generally not the case in either the US or the UK. This meant that there was a shortage of engineers in France. Even so the firm was a major recruiter of graduates. One of the biggest problems in the relationship was that the universities and CNRS were under-funded and they believed that industry was stacked with money.

The respondent's view was that the firm does not drive the direction of university research in France and that it would not be desirable for industry to do so. Its policy was to support long-term research in universities. It was thought that it was not appropriate for a CNRS laboratory or a university always to undertake research in conjunction with industry. The relationship with UK universities was different. The company had found them easy to work with and IPR was not a problem. The rights remained with the people who invented new ideas, except on joint research. The centre operated a strategy of institutionalizing links with universities in the US. For example with MIT it participated in industrial liaison programmes and had an industrial liaison officer in the university. This way the firm could obtain inside information about research and researchers. This was important because the overlap of interests with that university were considerable.

E. An ASICs manufacturer in Belgium belonging to a French telecoms firm

In a relationship that would not be possible in the UK or even in France, this firm had externalized a major part of its R&D activities to IMEC, using the laboratory as a third location of production. The relationship with IMEC was supported through the whole range of interactions possible under the laboratories' institutional set-up. They included (a) participation in international programmes, (b) recruitment, (c) consultancy, and (d) industrial residency. Most of the firm's engineers were recruited from either IMEC or KUL. In 1991, the firm had signed a special framework agreement so that the firm's engineers worked in IMEC. The intention was to use IMEC more as a prototype, pilot and development line. Although it was expensive to pay IMEC to undertake these activities, it would be even more expensive to disturb the firm's production line for research and development projects. Links with IMEC and the universities were increasingly important but were said to have improved competitiveness indirectly rather than directly.

Complex legal arrangements governed ownership of IPR between IMEC and the firm. Division of commercial rights and rights to principles arising from contract research were decided on a case-by-case basis. Where the work done by IMEC was more fundamental and could be re-used in other programmes, the arrangement was that the laboratory would be given the opportunity to take out a licence so that the technology could be re-used. In formal programmes the standard rules applied.

In sum, the general tendency for externalization, collaboration and networks to be increasing is exemplified by these case studies. All had extensive links with institutions in the home country and most participated in EU programmes. The main differences tended to be related to the construction of national innovation systems, particularly the kinds of institutions in the PSSB in particular technologies, the rules set on a case-by-case basis and the cost structures of interaction. For example, there is no UK equivalent to specialized institutions such as IMEC or CNRS or INRIA which means that firms have no choice but to seek out specialized expertise in a range of universities and national laboratories. While this broadened the scope of innovation filieres it meant that there are fewer possibilities for economies of scale in the use of resources. An important point emerges from the brief references to IPR which is that public programmes, particularly those organized by the EU, are essential to safeguarding universities' intellectual property. The responses from the UK sample suggest that cost is a more significant factor in firms' decisions on whether to interact than in France and Belgium and that it has been cost-effective to collaborate with a range of institutions. However, the balance seems to have shifted in favour of more work being undertaken with a smaller number of institutions.

3.7   The mediating role of trade and local industrial associations

Trade associations in this industry appear to have a minor role in organizing innovation filiere in this industry. In the UK, the ECIF had no interest in generating interaction between its members and the PSSB. In France, SYCEP was not a source of information about new technology for the one firm which was a member.

Other associations have treated fostering links as a minor objective, for example, the International Society for Hybrid Micro-electronics (ISHM). ISHM has three main chapters: in the US, Japan and Europe, with smaller local chapters. Its ability to facilitate links is limited for two reasons. First, it tends to be dominated by industry and therefore is not institutionalized into academic networks. Secondly, as the market for hybrids was not increasing, firms are less likely to adopt an aggressive (costly) strategy towards integrating its members into formal relationships with the science base. The nearest equivalent to innovation clubs in flow measurement is a GaAs consortium which meets once a year by invitation only, in a different locations. Two firms (one in the UK and one in France) belonged to ISHM but neither reported that any support for working with universities had been given by the organization.

4.   Benefits and perceptions of universities

4.1   Benefits

The main advantages of interaction in this industry were in some respects similar to those in the flow measurement industry. As in that industry, there was evidence that the PSSB was involved in short-term problem solving but this was offset by a far greater commitment to long-term projects co-funded by national and EU programmes. Through both means, links with universities and national laboratories were seen as essential to firms' competitive positions. The strategies varied between firms and depended on the nature of the institutions with which they were working. For some, contact was quite clearly externalization at a sophisticated level – for example the Belgian firm working with IMEC and Japanese research centre in the UK both said that because work was being undertaken in universities on its behalf, it did not have to take on staff. The universities' importance as a sources of new employees provided the major contrast between this and the flow measurement industry.

4.2   Problems of interaction

Some problems were common to all three countries and others which were country specific. General issues were related to under-funding of research in both universities and national laboratories, timescales, IPR and for those without links, a lack of information. Universities were seen to be operating at a more leisurely pace than industry. A lack of information was mentioned as a factor which inhibited the formation of links by both smaller firms and branch plants in the UK and France. This was not mentioned in Belgium which has a small and well net-worked electronics community.

In the UK, all of the sample which collaborated with the PSSB were critical of some aspect of the working relationship. The cost of funding research in the PSSB was an important issue. One major UK firm said that cost had recently become more of a factor when decisions were being taken on what and where to pay for research. The problem of fragmentation of university research, and thus the need to involve several universities in projects, was mentioned by three firms. The comments on national laboratories were not about the quality of work which was described as excellent but were to do with what one firm described as their 'commercial immaturity'. Generally how quickly the work got done and the quality of the relationship depended on individuals. There was also a concern about the impact that the new status would have on the quality of work in what was then RSRE Malvern.

In the French sample, links were less institutionalized through both formal means and through informal networks than those in the UK and Belgium. Contacts with universities and engineering schools were often *ad hoc* and not repeated. The sample had more complaints than firms in the UK and Belgium. Universities were criticized for conflicting reasons. Firms complained on the one hand that there was insufficient expertise in particular areas of engineering and that universities were not good at solving real life problems, and on the other that they had the expertise but did not spend enough time finding out what industry needed.

In Belgium, firms identified the general problem of under-funding of university research as limiting co-operation with industry. The competition between universities for funds and the resulting lack of collaboration meant that resources which could have been shared were being wasted because they were under-used. Firms were not prepared to provide the extra funding to bring together groups in different universities. Firms in the Belgian sample which currently had, or previously had, close links with IMEC had a number of criticisms. The first was the culture of the laboratory. It was felt that there had to be more of a move to an industrial from an academic mentality. In the opinion of the respondent from case study E above, IMEC should be driven by industry, particularly local industry and should act as a subcontractor. The second criticism was that a disproportionate amount of funding for IMEC was at the expense of the universities. In a move to ease tensions and increase co-operation, since 1992 10 per cent of IMEC's research budget has to be spent in the universities. The third comment was that IMEC was not selective enough in the areas in which it specialized. This put its scientists in competition with university departments which had built up expertise over many years.

5. Summary of major features and conclusions

The norm for this sample is that firms in each country have some form of interaction with universities and nearly half had links with national laboratories. The evidence showed that the main regulators of the structure of the filiere in terms of whether interaction was increasing or decreasing, duration of links, which organizations were participants and where the organizations with which they had links were located, were firms' position with company hierarchy, recruitment, participation in public programmes, cost (in the UK) rather than size or ownership or the perceptions of the role of universities and national laboratories in the innovation process.

The evidence showed that research centres in each country had the most extensive links. The companies which did not have links were mostly the production branches of MNCs. The research centres were an important influence on the construction of innovation filieres. They acted as co-ordinators for certain activities within the firm while inhibiting participation by the engineering operations of their companies. At the same time there was evidence that most branch plants were intending to or were already seeking technical support from universities. This was a result of competitive pressures. In product markets where technological change was slower and less radical competition from cheaper technologies and price competition, firms anticipated that universities could help with developments in engineering. A barrier to interaction was that industrialists often did not know whom to contact in the PSSB.

While branch plants were a new source of demand, there were also indications that firms with well-established links, particularly in the UK and Belgium, were increasingly looking to fund more applied rather than speculative projects. For some this meant long-term R&D collaborations were being displaced by shorter-term projects. For others it involved interactions with fewer universities.

The spatial pattern of interaction showed two dominant trends. The first was that most linkages were occurring within the national space. There was limited evidence of proximity being a factor in whether links were local in the UK but there did seem to be an association in France reflecting the large distances between centres of excellence. In such a small country such as Belgium, the electronics technological community is not anchored to one location, although IMEC is a pivotal organization. The other trend was that there was an increase in formalized interaction with universities in other countries in Europe, and in the UK particularly, with ones in the US. This was not size-dependent. Smaller firms were equally as likely to have links with foreign institutions as larger ones.

A further regulator of interaction is that of the ownership of intellectual property rights. The evidence from this sample is that on the whole universities tended to give up rights to IP. This factor, and pressure caused by decreasing willingness of larger firms to fund open-ended research, meant the mode of interaction was changing. This industry funded less speculative research as competition increased. These forces changed both the practice of in-house R&D and by extension the pattern of contact with the PSSB.

In this sample, unlike in the flow measuring industry, ownership was less of a regulator of interaction except in the case of US-owned firms. In both this and the flow measurement industry, those with US parent-firms had less freedom to chose their own R&D activities than branch plants of firms from other countries. Take-over *per se* had not been a universal direct cause of decline in externalization and collaboration. In some cases, for example in the acquisition of a British firm by one from Canada, R&D investment and interaction with the PSSB had increased. In contrast, merger between two UK firms had reduced the overall level of interaction because R&D activities which were duplicated were consolidated at one site.

Recruitment from the PSSB was evident in the most innovative firms, particularly in the UK and Belgium, providing both expertise and access to national and international university networks. In France this means of strengthening the filiere was limited by the shortage of engineering graduates.

Participation in national and international programmes was a major regulator of power exercised within relationships, of the social qualities of interaction, the duration of links and the spatial construction of the electronic components filieres. The sets of rules which determined access and control of intellectual property allowed more control by universities of their intellectual property than on contract research. Most programmes, for example those which supported doctoral students such as CASE awards and those which funded basic research such as ESPRIT, lasted for three years. From the comments made by respondents it appears that these formal programmes have created networks, a culture of collaboration and co-operation and have institutionalized links. Collectively they contribute to an innovation filiere constructed from diversely located elements.

The survey showed that there was little difference in the stated value of universities to manufacturers and research centres in the three countries. All companies with links said that they were important to current and future competitiveness. Universities provided complementary expertise, at a basic scientific/theoretical level, they undertook longer term research, ahead of time of industry needs, thus reducing industry's risks of investment in areas where there might be no payback and were frequently used for problem solving, especially in the UK. The possible adverse implications of this are that local and non-local knowledge was not being exploited which may have contributed to more productive methods of innovation.

8

The Views of Universities and National Laboratories: the Electronic Components Industry

Introduction

In Chapter 7 the focus was on the use of universities and national laboratories by the electronic components industry and the various means by which links were constructed. The main message from that chapter was that there were distinct differences in expectations between what industry wanted and the way that PSSB institutions responded. Here the opposite (or complementary) point of view is presented. The main theme which emerges is that there is a strong sense of identity with the electronics industry in all three countries but at the same time there is considerable frustration at the distortion of the balance between short-term and long-term projects as the necessity to provide a service to industry takes priority over 'blue sky' research.

The chapter begins by introducing the characteristics of the university and national laboratory samples (specializations, size of groups, funding arrangements, and so on). The following section records different aspects of the interaction with industry. The purpose is to examine which factors regulate the form and outcomes of these linkages. The final section summarizes the main findings and draws some conclusions.

1.   The sample

1.1   Sample profile

In total, 14 institutions were visited in the three countries (Table 8.1). These comprised ten universities, five in the UK, one in France and

Table 8.1   Sample PSSB institutions with research in electronics in the UK, France and Belgium

Department	Staff	Specializations	Research Funding
UK			
1. Electrical and electronic engineering	Group: 4 academics plus 20 technicians	Silicon on insulators, unusual electrical properties of devices, developing and understanding novel device structures	MoD, inc. studentships, SERC, LINK, larger UK firms
2. Physics	Group specialising in properties of semiconductors, 1 academic plus several permanent SERC funded RAs	3.5 structures, electrical properties, growing structures	LINK, CASE students, SERC – research assistants, fellowships and a rolling grant
3. Programming research group, specialist group (Computing Laboratory)	One full time plus RAs and doctoral students	Parallel computing	DTI/SERC ESPRIT, UK and French-Italian firms in the UK
4. Microfabrication facility (electronic engineering)	10 research fellows, 10 technicians and 9/10 PhDs	Micro-electronics – integrated circuits, structures on silicon	SERC, LINK, CASE students, UK and US firms with UK operations
5. Physics	4 industrial staff	Semiconductor multi-layer physics for devices, infra-red detection and communications	Larger UK firms, SERC, LINK, CASE students
6. Physics	20 academics, 27 post-docs, 36 research students and 12 technicians	Interfaces in semiconductor structures, laser physics	SERC, DTI/SERC LINK, ALVEY, MoD, CASE students, UK, some small firms, Japanese, Canadian and US industry
7. DRA Malvern	25 physicists	New device functions	ESPRIT, MoD, LINK, CASE

Table 8.1 Continued

Department	Staff	Specializations	Research Funding
8. Harwell	2 groups i) silicon radiation detection ii) semiconductor research	Semiconductor research fabrication facility and an ion-implantation facility. i. Diamond thin film, ii. silicon germanium	BRITE, UK industry, Swiss industry
France			
9. ISEN	65/70 staff working on microelectronics	Microelectronics	JESSI, ESPRIT Training subsidy from industry, the state, CNRS, small companies
10. LETI per cent	1000	Micro-electronics, opto-electronics, systems, detection	10 per cent CEA, 90 industry, mainly French defence industries, studentships, ESPRIT, JESSI, BRITE, RACE and EUREKA projects
Belgium			
11. Electronics and Metrology	90 (10 permanent, of which 7–8 academic staff). Thin film component group11 staff	i) thin film components ii) digital systems iii) signal processing e.g. speech recognition	75 per cent research income project based (industry, inc. US DoD, national and Flanders funding institutes, NSF studentship) 25 per cent university funding RACE
12. Electronics systems automation technology (Department of Electrical Engineering)	18 full time, 9 Research associates funded by National Fund of Scientific Research	Integrated circuits and sensors,	50 per cent industry (Belgium, France, Netherlands, Japan) 50 per cent national institutes e.g. IWOLN

Table 8.1   Continued

13. Applied physics	n/a	Opto-electronics, and fibre optics	French Industry, ESPRIT
14. IMEC	390	ASICs	Annual budget c 1.3 Billion BF. (1992) Flanders government 60%, contract research from industry (about half of other sources), 31.4% of the total contract research income originated from Flemish industry ESA, ESPRIT, JESSI, French industry. Local and federal government departments, studentships

three in Belgium, and four national laboratories, two in the UK and one each in France and Belgium. All but two university departments in the UK were physics departments, the others being one of the two Microfabrication facilities and a computing group. In France and Belgium in all but one case, the samples comprised electronics research groups in engineering departments. The one exception was an applied physics group in a Belgian university. Specializations included the properties of semiconductors, ASICs, fibre optics and parallel computing. Each of the national laboratories in France (LETI) and Belgium (IMEC) specialized in electronics. This was one of RSRE Malvern's two specializations, the other being command information systems.

1.2   Research funding

An indicator of the kinds of research involving industry in which departments were engaged and hence their degree of integration within *innovation* filieres was the sources of funding they received. Most were involved in industrial innovation activities through direct industrial funding – most also had funding that was not directly funded by industry – but national laboratories were more likely than universities to be part of European-wide R&D programmes. This suggested that national laboratories had already moved farther towards integration with industry than the universities.

There were some common patterns and some national differences. All of the sample, both universities and national laboratories, were engaged in collaborations with industry; all but one of the university sample, the applied physics group in Belgium, at the time of the study held national research council grants; most universities and national laboratory departments were involved in joint projects with industry under national innovation support schemes. An important difference between the UK and Belgium was in the number of national studentship awards. While five of the six universities in the UK but only one of the four Belgian departments held studentships, all but Harwell of the national laboratories had studentships.

All of the national laboratories had grants from EU programmes such as ESPRIT, BRITE and JESSI, but of the university groups, only one of six the UK universities, the one French and two of the three Belgian universities, had European awards. One UK department recognized the need to be in European programmes but its applications had been unsuccessful. The Harwell group had worked with IMEC on an ESPRIT Basic Research programme.

While national and European programmes including CASE studentships and ESPRIT projects supported longer-term projects, industry projects tended to be much more short term. Table 8.2 shows that most UK and Belgian departments were undertaking short-term applied projects and that their role within innovation filieres was that of a subcontractor to industry, undertaking externalized projects as a substitute for in-house activity. One UK department, the computing group, had been able buck this trend. It was able to accept industrial funding on the basis of either whether the group could have an impact on the direction of industrial research and change company thinking, or where the research had long-term significance.

Other forms of industrial support universities received included donations of equipment, sometimes in return for services such as testing. The value varied in importance between departments. For example, donations of equipment were a very significant contribution to the income of the UK Microfabrication Facility. It had received some £1 million of hardware and software, which amounted to about 12.5 per cent of its total investment. In return, the Facility acted as a test site for a US company and for another it tested software and provided feedback on the results.

Defence ministries directly funded research in two UK and one Belgian universities and in two of the four national laboratories, Malvern and LETI, which were central components of their respective national defence innovation system. Indeed, LETI's biggest customer is the Ministry of Defence. In Belgium, funding was from the US defence budget rather than the national defence department. Both Harwell and IMEC were explicitly prohibited from being funded directly by defence ministries. The policy of the Flemish Government is that IMEC should not be engaged on defence work. However, they were still part of the defence system because they collaborated with companies working on defence projects.

The theme of competition between universities in Belgium for funding which was mentioned in Chapter 7 appeared again in the answer to the question about problems faced by academics. One UK respondent commented on the difficulties caused by the fragmentation of research in the UK. While 12 groups might compete against each other in the UK for European funds, IMEC could get large awards because it was a centre of expertise and therefore occupied a central position within the Belgian and European electronic components innovation filiere. None of the UK universities had sufficient breadth of research to be included on some of the big European contracts. The respondent suggested that a solution might be a distributed micro-electronics centre – a distributed IMEC – which used the expertise and facilities of the country's major departments. Moreover, better co-ordination of work on research council funded projects would bring together people working on existing grants into a collaborative frame-work. This would avoid duplication and generate a stronger commu-nity. It would also be a means of linking academic-only programmes into national and European networks.

1.3   Geographical patterns

The research showed that there were national differences in geographical orientation of research linkages. Four interwoven strands contribute to the differences in the geographical construction of innovation filieres.

A. The location of the source of direct funding from industry

In the UK and France, contracts from industry mostly came from firms located within the country. In contrast, all of the Belgian sample were paid by companies based elsewhere. The majority of UK university and national laboratory departments worked primarily with larger British companies and the UK-based operations of foreign-owned firms, particularly from North America. The reliance on European funding in computer software had re-configured research patterns for the parallel computing group. Whereas it had previously collaborated with US partners, it had begun to collaborate with German universities. In this area of research, French universities were not part of this newly created European research community. For LETI, the transfer of technology to industry outside France, particularly to Japan and the Far East, had to be authorized by the relevant ministry. It was in principle less difficult to license research results to US industry but in practice there was little in the way of co-operation with US firms. Contracts from foreign industry in both universities and IMEC in Belgium came mainly from within Europe, particularly neighbouring countries such as France and the Netherlands. One department was working on a project for a Japanese company.

B. The spatially integrating function of EU programmes

Internationalization of R&D through long-term collaborations draws together expertise from diverse locations, broadening the networks which are the 'social glue' of innovation filieres. While technological advance is the prime incentive for collaboration, the funding opportunities were high priorities not least because of the externality effects of the extra income as well as the leverage brought by association with other international players. Success at European and broader spatial scales has had a reinforcing effect on geographical concentration of expertise. The expanding role of IMEC is the best example of this process.

C. Proximity

The local accumulation of the number of firms in contact with IMEC and LETI arose from similar processes but operating for different lengths of time. Both were funded to participate in technology transfer activities both locally and non-locally, and both contributed to the expansion of the electronics industry in their region (see below). The strong local focus innovation filiere which had IMEC as its centre is demonstrated by the statistic that in 1991 nearly a third of IMEC's income came from industry in Flanders. Similarly LETI had close links with the electronics industry in Grenoble, with both large and small firms. LETI's relationships with smaller firms in the region include:

- giving technological advice
- identifying when the solution to the problems of SMEs can be found within CEA

- helping SMEs identify mechanisms to fund innovation
- directly part funding innovation, for example can share up to 50 per cent of the expenses on a project which has some risk and helping the small company get help for the other 50 per cent from organizations including CRITT and ANVAR
- innovation clubs.

Both CRITT and ANVAR rely on public funding to underpin their technological positions with local, national and international filieres. IMEC is reliant on funding from the regional government, while funding from the CEA, albeit only 10 per cent of income, and the heavy subsidies to French industry both enabled LETI to maintain its research activities. The quality of the infrastructure through major capital investments were of crucial importance in attracting industry funding in both institutions.

D. The strategic role played institutions and individuals in attracting and retaining industry to particular locations

The examples of IMEC and a UK academic illustrate this point.

IMEC   Part of IMEC's recent remit has been to attract foreign investment. The possibility of working with IMEC is used by FLOC, the Flemish Government Council, as an attraction to firms from Japan, the US, Scandinavia and Europe to locate in Flanders. This strategy has met with some success. Alcatel had moved the majority of its R&D into Flanders in order to be near to IMEC and Philips had set up a unit in Leuven to work on ASIC design with its engineers. Some research departments of Philips have remained in Belgium because of their links with IMEC.

A UK academic   The only UK example where an individual academic had been involved in inward investment was where an academic had been acting on behalf of the Welsh Development Agency. In 1985 a professor had visited firms in Japan which had manufacturing units in Wales. His mission was to assure the companies that higher education in Wales was listening to the training, research and education needs of Japanese companies.

1.4   Changes over time

In the introductory chapter, a filiere was defined as a 'structure in process' because at every stage it is subject to pressures for change resulting from changes in technology, market structure or government regulation (Preston 1989). Of these, from the point of view from the

PSSB, the two most important were regulation and market structure. As has been demonstrated earlier and is further illustrated below, regulation through decisions on allocations of public funding to universities and the status of national laboratories, particularly in the UK, was the major factor in shaping the industry-PSSB filiere.

For the UK national laboratories the changes in status and emphasis which brought a much more commercial focus was the most significant factor in the changing nature of relationships with industry. For the group in RSRE Malvern three immediate affects of the change to agency status on relationships with industry were to re-evaluate research strategy, to increase efficiency and to force the laboratory into a competitive rather than a collaborative situation with industry. The last of these meant competing with the research centres of GEC in the UK and with those of Siemens and Thomson on mainland Europe. The overall effect was to drastically reduce the number of defence related links with industry. On the other hand, there was more freedom to seek funds for collaborative projects on non-defence related work than before. The respondent was offered more business than his group could cope with from companies such as Siemens.

The Harwell Facility originally existed to supply detectors to the radiation industry. The effect of change in laboratory status was that it had started to identify applications of new technology, for example photodiodes and sensors as devices to measure pressure acceleration, and to look for contracts with industry both UK and in Europe. The group's commercial activities were possible because it had a silicon processing capability developed from its work on detectors and because for some time it had been allowed to manufacture in non-nuclear areas on site. However, its opportunities were limited by operating in a small market which included the nuclear industry, the MoD, universities and hospitals.

Changes to UK laboratories had also had an impact on the universities. The reduction in the level of 'blue skies' work meant less of an overlap of interest and therefore possible sources of funding and of samples of materials. In France, the increasingly commercial orientation of LETI meant that researchers faced the problems of being obliged to go in too many directions and of having to undertake short-term research. The respondent said that companies wanted research results which could be translated into products. This created a conflict between short- and long-term goals. Major industrial customers such as SGS-Thomson had a major influence on the direction of research in the laboratory. IMEC had evolved as a result of changes in the policy of the Flanders government towards the role of the laboratory. At the outset, IMEC was primarily a research organization undertaking long-term research. By 1992 it had responsibilities for helping SMEs in Flanders (see below) and, like the other laboratories and the universities, was expected to undertake short-term research.

An example of the impact of changes in market structure is the decline in industrial research, particularly in silicon in major UK companies. For one group this meant that it was becoming difficult to find high-level researchers in industry who would drive the research. The respondent said: 'Over the years the group has worked with most of the British semiconductor manufacturers. It changes with time as they go in and out of business.' The effects had been threefold:

a)   to reduce the number of CASE studentships co-funded by industry;
b)   to reorientate research away from areas with potential industrial application to focusing on fundamental research funded by research council or DTI grants;
c)   to increase collaboration with research groups in foreign MNCs.

For another UK group, current grants for work on opto-electronics had supported research without pressure from industry, and had delayed the problem of the decrease in funding for research on silicon. This group had fared better than some laboratories which had been harder hit as the reduction in SERC funding for silicon research had been particularly severe. The respondent said that he, 'sends students on all sorts of academic detours, building up experience in the way that industry would not be interested in'.

2.   Interaction

2.1   Forms of assistance

Academics were asked to describe the forms of assistance which their group or department offered to industry, choosing from the same list of possibilities which the firms had been given (Table 8.2). The table also lists the current projects involving industrial partners. This shows that there were national differences between the UK and Belgium in the forms of assistance provided to industry. All of the UK but none of the Belgian respondents said that they offered basic scientific assistance. The responses of both university samples and of LETI indicated the prevalence of applied research in their activities and therefore their intimate engagement in industrial research. Four UK departments said that they solved development problems and three assisted in product

Table 8.2   Sample departments' relationships with industry

Department	Forms of interaction with industry
UK	
1. Electrical and electronic engineering	Basic scientific knowledge, new generic technology, assistance in new product development, occasionally new testing regimes, highly skilled PhD personnel. The group currently had two projects: i. Silicon on insulator collaborative project MoD funded – 1 academic, 3 RAs and 2 research students. This used to be SERC funded. RAs are funded from sources other than the MoD contract. Under IEATP. ii. 1.5 members of staff, plus 2 RAs. This includes two British firms and one Canadian. Other forms of contact (i) secondments, both to and from industry for short periods. This was a frequent activity involving firms such as STC, Siemens, GEC. (ii) an industrial fellowship (in the department but not that group), (iii) specialist processing for industrial use and (iv)recruitment of PhDs.
2. Physics	Basic scientific knowledge, some short-term problem solving e.g. characterization – one SERC funded joint research programme, rolling grant.
3. Programming research group	Basic scientific knowledge, new generic technology development, assistance in new product development, solving production problems, improving production processes, routine testing and validation services, and adaptation of technology to deal with local conditions. Most work 5 years ahead of the market. Stopped undertaking short term industrial projects in the 1980s. The group had three industrial projects: (i) subcontractor on an EU programme, gaining about 2% of the grant. (ii) project under the Basic Research Action section of ESPRIT. (iii) PUMA project – objective to develop a family of transputer based components providing low latency communications. MSc students on industrial bursaries.
4. Electronics engineering/ Microfabrication facility	Basic scientific knowledge, new generic technology, assistance in new product development, adaptation of technology to deal with local conditions, solving product development problems, and technical reports for individual firms, processing, testing facilities, target research of commercial interest. Provides a service role for academia. Projects: 2 LINK projects programme initiated under JOERS – for spatial light modulator technology with BNRE and GEC. CASE award student funded by Olin-Ciba-Geigy. Department had 30 MScs a year on industrial bursaries for example from INMOS, BP and Smiths Associates.

Table 8.2 Continued

Department	Forms of interaction with industry
5. Physics	Industry pays for longer-term research which could not be done in-house, company policy was to make maximum use of public facilities, provides basic scientific knowledge, prototype development, developing testing routines.
6. Physics	Basic scientific knowledge, assistance in new product development, and solving product development problems Projects: i. A directors of the research centre of a UK telecommunications firm is a visiting fellow working on opto-electronics. Paid contract to develop quantum wire lasers, plus 2/3 CASE students. ii. With a small local firm. The department makes laser structures for materials for the company and has many projects with it including a teaching company scheme and CASE awards and 2 LINK grants. iii. A LINK project. 1990–1992, on opto-electronic integrated circuits – involving a major telecoms firm, a small materials company and one CASE student funded by the telecoms company.
7. DRA Malvern	Civil applications of military funded research into new device functions bulk of the research effort is applied research, aimed at 5–10 years ahead of current military needs. Spectrum of research from innovation through to systems testing and engineering aspects. Expertise in GaAs and liquid crystal display technology.
8. Harwell	Strong analytical capacity, problem solving, processing.
France	
9. ISEN	Not usually direct industry funding, more often in French or EC funded consortia, with partners in the UK, in France with Thomson, and the French navy.
10. LETI	Generic technology and help with improving production processes. Its only activities in passive components was on special techniques for interconnections of ICs. No facilities for research on basic components such as resistors and capacitors. Industrial residents, licensing and patents, research programmes driven by industry especially French industry, support for SMEs Research funding (i) under contract with private industry and (ii) by government departments.

Table 8.2 Continued

Department	Forms of interaction with industry
Belgium	
11. Electronics and Metrology	R&D for electronic components made by thin film technology. Applied research. Thin film group, industrial funding direct or in joint projects with industry fund 10 of 11 staff. Industrial support from Finland, Japan, Canada, Spain. Some scientific research, most are short-term projects. 8 projects each funded differently: i. A research project for fundamental research funded by the Flanders Ministry of Science. This had no industry applications. ii. EC project – RACE – the group was a subcontractor to the head contractor, a Finnish firm, directed towards developing large area displays. iii. A researcher was funded by a renewable six month industrial contract from a US SME which has a USA Defence Army Research Project. iv. The Institute for Science in Agriculture and Industry grants for PhD students. v. A graduate student researcher paid by the national science foundation to undertake pure research. Foundation was due to be administered at the regional level. vi. Funded on a government initiative for parallel computing. vii. A researcher paid on an industrial project with MIETEC. Half of the costs are paid for by regional funding body. viii. A researcher from China, funded on departmental grant using money gained from an industrial contract.
12. Electronics systems automation and technology	Adaptation of technology to local conditions, new generic technology development. Ideal projects last 2/3 years, may have to solve specific problems as part of project.
13. Applied physics	Application from a discovery in the department, development of product in conjunction with Harwell, and a French telecoms company
14. IMEC	Contract research 5–10 years ahead of industry's needs especially in integrated circuit computer aided design tools. Training, support for SMEs. Projects: i. EC In 1984 there were 60 EC programmes, by the end of 1990, 46 were still going. Industrial partnerships involve more than 2 partners, from more than one country – supra-national consortia ii. European bi-lateral contracts outside EC framework iii. Regional/national bi-lateral contracts – 2 ESA

development. LETI, two of the UK universities and one of the Belgian universities assisted in the development of new generic technology and adaptation of technology to local conditions. In Belgium, each department attempted to 'ring-fence' research that had no immediate benefit to industry. For example all academic staff in the Department of Electronic Systems, Automation and Technology spent some of their time on fundamental research in order keep some selectivity in the kind of projects which were undertaken.

2.3   Initiation

Personal contact was the most significant factor in establishing links in each country. Formal mechanisms such as industrial liaison units were infrequently the means of matching firms to individuals. It was common for UK academics to approach industry with an idea. Other ways in which contact had been made by industry or vice versa were through made through research literature, conferences and CASE awards. These provided the means of identifying interests on both sides.

The close relationship between industry and the PSSB is illustrated by an example from the UK, a university group specializing in advanced materials. The decision on which material systems to work often depended on whether it might also be interesting to industry. The respondent said that much of what academics were doing in this area was determined by the skills of industrial materials scientists, which often takes years and millions of pounds spent on developing structures to acquire.

2.4   Movement of personnel

The movement of personnel between the PSSB and industry did not take the same direction in each country. There were three main differences. First, in the UK, flows tended to be from industry and from national laboratories into universities. A feature of the UK system in the early 1990s was the regular career progression of researchers in industry moving to posts in universities. It is common for university academics to have worked in industry and to retain strong contacts. The impact of the decrease in basic research associated with the more commercial orientation of national laboratories had been for national laboratory personnel from both Harwell and Malvern to seek more freedom in research in universities (see also Lawton Smith 1997).

The second was between the UK on one hand and France and Belgian institutions on the other in that the system of industrial residency was well established in both LETI and IMEC but not in the

UK national laboratories or universities. In the UK temporary mobility took the form of secondments, industrial fellowships and in only one case, industrial residency. The instance of industrial residency in the UK involved a group industrial researchers installed as industrial residents in the physics department of one of the UK's premier universities by a major UK electronics firm. The respondent, who was both an academic and an industrial research scientist, explained that the secondment was the result of a change of policy in the firm and would not have happened ten years earlier. The objective was to make maximum use of public facilities, including expensive equipment because 'The cost of research is going up relentlessly.'

In France, mobility between LETI and industry consisted of four categories:

(i)   people leaving to go to industrial partners such as Thomson, Peugeot and local electronics firms;
(ii) industrial residents;
(iii) people either forming or being recruited by start-up companies;
(iv) people returning to LETI from industry.

About a fifth of LETI's population were temporary residents of various kinds mostly graduate students. The second largest group was 73 industrial residents based full time at LETI. In 1992, there were more than 15 people from SGS-Thomson based in the laboratory working on a joint programme on CMOS technology which was funded and driven by the company.

Links with industry generated by movement of personnel were strongest in IMEC. The laboratory has a general turnover of about 15 per cent a year; some 69 per cent of personnel leave because their contracts have ended. Most are recruited by industry. The system of temporary contracts was a deliberate policy. The policy makers in the Flanders government believed that because the future for industry lies with its skilled people, it was necessary to have an institution capable of raising the level of skills to that of industry, particularly at the level of post-graduate education. The number of industrial residents at IMEC changed over time. For example, at the end of 1992 there was a large group of industrial residents from Philips.

The third difference was mobility in the form of entrepreneurship. A formal contribution to mobility not developed in UK national laboratories but which was a mission of both LETI and IMEC is the formation of spin-outs by scientists and engineers. By 1994 LETI had been the originator of some 15, mostly small, firms. The largest to date was EFCIS, formed in 1972, which had created some 80 jobs. By 1993 IMEC had been the source of seven new firms (Lawton Smith 1997).

2.5 Benefits to interaction

The most direct benefit from working with industry in all countries, which was the same as in the flow measurement industry, was income to support both academic and student research. Other benefits included example of real life problems for use in teaching, use of firms' facilities such as clean rooms and the mutual benefit of gained from characterization of semiconductor structures. A UK academic summed up the department's attitude towards working with industry: 'We recognize that our future is tied up with industry, we don't want to be diverging from industry, we want to be working on similar sorts of problems and have them be interested in what we do.' The benefits to interaction to a greater extent were based on mutual interests. The academics provided new knowledge and used their expertise to help industry with practical problems. Firms provided funding that would sustain research in particular directions and access to facilities and materials that departments were unable to provide.

2.6 Problems with interaction

While there appears to be evidence that groups in both universities and national laboratories have been successful in obtaining external research income, the majority of respondents said that they were under-funded. This was the basis of problems faced by universities and were compounded by the expectations of industry about what universities and national laboratories would deliver.

Reductions in state funding and pressure to obtain money from industry had meant that academics were less able to concentrate on more fundamental research than was required by industry in the short to medium term. For one department the reductions in capital investment in key technologies such as silicon had the consequence that it was tending to concentrate on basic research, which was cheaper, while its traditional partners in industry had focused on development. This might have implied a complementary relationship but it actually meant that undergraduates and graduates were not trained on the latest processes and that possibilities for joint research were limited. Another group was vulnerable to industry's propensity to change their research interests over a short period of time and refusal to indicate what they would be doing in the future.

There was a general recognition of the dangers of undertaking short-term research. The Head of a UK department said that he felt that his department as a whole had too much short-term research. On the other hand it was necessary to take on industry projects in order to fund increases in the department's research portfolio. Another was very careful not to get caught on too many short-term projects where the results could not be published. The respondent thought that generally there was a danger of some departments doing too much fire-fighting for industry.

Another issue which emerged in the UK was the conflict between researchers and the university contracts office over the level of over-heads charged to industry, hence the cost of the project to firms. In one case, the administration wanted high overheads to maximize income. The academic viewpoint was that this was unrealistic. If work was done for industry without academics being able to publish, then industry should pay the whole cost. If it was publishable, then the work would not have been much different to what the department would have been doing anyway. On that basis the overhead would be approximately 40 per cent – the same as was charged on research council work. On the other hand, if the work was very industry-orientated overheads might be as high as 150 per cent.

The most important problem identified by the French university mirrored the comments made by the respondent in one of the research centres. His department suffered from a lack of industrial funding. Its main activity was teaching, which produced good students valuable to industry.

In Belgium, the lack of funding for research was the most serious problem facing university departments. The academics, like the indus-trialists, were critical of what they saw as the disproportionate amount of research funding which went into IMEC. The shortage of research funding increased the degree of dependency on industry which meant that there was less choice about which projects could be undertaken. One academic said, 'There is never money enough to do work by your-self without interference of industry. The question is, what is the minimum amount needed to be able to do so?'

One Belgian department had learned to be creative in finding arrangements whereby industrial money could be used to fund longer-term research. Firms paid for their research and they asked for results for about half of that money. The other half was used for more fundamental research on projects which could produce specific results. For this they needed information from industry in order to be able to propose future research topics. One of the reasons it worked quite well was that the group did not do research that might have a result in ten years time. The group responded to industry's timescales. If it was a software problem, it would try to work a timescale of between four to six years, if it was an electronics problem, the timescale would be from two to four years. The respondent stressed that the links, although applied, were not problem solving. However, when a project with a firm involved developing something new for two to three years later, it would be difficult to refuse to help the sponsor with some specific problems.

The complaints of the universities were matched by those in the national laboratories. The Malvern respondent said that industry was conservative in its attitude, and took a short-term view, wanting to pay for shorter contracts. Moreover, in its new role, Malvern was competing with universities which had lower overheads and more flexibility. His group which was small and operating in niche markets faced the danger of being squeezed as it entered a competitive market for funding from both universities and industry.

Problems of what in academic terms constituted too close a relationship with industry are illustrated by experiences from the UK. One concerns the fortunes of a group within a university, the other those of an individual.

The first is the case of the Microelectronics Fabrication Facility. Its relationship with industry had become out of step with the prevailing research council policy and as a result had lost its rolling grant. The respondent's view was that SERC's attitude appeared to be that because the indigenous semiconductor industry (for example, Plessey, GEC, Ferranti) had reached a point of extinction, the only way forward for universities was old-fashioned blue sky research. This was at odds with the Facility's strategy of undertaking applied research and problem solving for industry. Its activities included silicon processing for customers such as Philips and Plessey. A contributory and related factor to SERC's decision not to renew funding was failure to publish. Publication rights often were held by firms. There had only been a few occasions where the names of people from the Facility had been on papers published by industry. This was the cause of the problem in the second example. In this case a member of staff was unsuccessful in his application for promotion because he had been undertaking too much industrial research and was then not able to publish.

2.7 The formal interface, IPR arrangements and publication

A key means of regulating the way innovation filieres function is through the criteria on which PSSB institutions accept research funding from industry. The degree to which information is controlled or flows freely through networks within 'technological communities' serves either to segment or integrate their components. A factor frequently mentioned as determining the nature of the relationship in all three countries was IPR arrangements on industrial contracts. The general position was that universities were not in a position to hold onto IPR. One UK physicist said, 'Industry takes universities to the cleaners.'

In reality this was not a problem because in his experience, a) academics have no ambitions to be millionaires, and b) much university work did not have IPR value.

For another group there were no constraints on which industrial projects were undertaken as long as the work was of scientific interest, was open and publishable. This was critical for research students, as for academic staff, since academic success depends on publication. One Belgian group had reached agreement with industry that the firm had the IPR for one year after the end of the project, but retained the right to publish the results of their contribution to the project. Sometimes industry required a delay in publication.

The national laboratories varied in their strategies towards ownership of intellectual property. The Malvern group as a contractor of research, did not tie up IPR rights to the work they funded in universities in the belief that universities should be doing longer-term research and publishing what they saw as appropriate. That situation could change as a result of the move to a more commercial footing.

LETI's approach to the ownership of intellectual property was frequently unpopular with industry. The laboratory's agreements with industry stated that all the know-how that both pre-existed and was developed in the course of research programmes belonged to LETI. If a company wanted to industrialize the results of this research they would have to license it from LETI. Sometimes a company would pay for 100 per cent of a research programme and then find that it had no rights to the IPR. The reason for this was that the firm was only paying for adaptation of the technology to the specific problem. Where possible LETI would offer a non-exclusive licence. This form of arrangement was to encourage exploitation of the basic technology. This policy had caused problems in EU consortia because partners were industrial companies which wanted to collaborate on the basis of ownership of the rights to technological advances.

2.7   Consortia and clubs

The two nuclear laboratories, Harwell and LETI, operate a number of collaborative clubs in electronics, for example one on optical codings at LETI. The difference between them is the basis on which they operate. At Harwell, members are mainly large firms and were seen as a means of disseminating information between equal partners. At LETI there was more of a 'public service' element. Small firms were encouraged to become members. SMEs paid an entrance fee and they had the right to come once or twice a year to a one day meeting. They also had state-of-the-art reports on technological developments. Clubs were not a very profitable activity, being more of a service than a money earner.

2.8   Committees

An important regulator of changing structure of innovation filiere is the means by which power and influence are exercized over the allocation of funds for both research in the PSSB and subsidizing industry's access to that research. Membership of committees where the allocation of funds is decided is therefore a critical 'esteem' indicator, that is those people whom the research councils trust to decide what kinds of research will be supported and the criteria for access. This appeared to be much more of an issue in the UK than in France and Belgium. The only academic who said that he was involved in shaping the direction of research was one of the Belgian sample who was on the board of IMEC.

Several of UK respondents, both academics and laboratory researchers, were members of research council and joint committees. One academic said that membership of SERC increased the department's chances of getting money and that industry was more friendly to departments with academics on committees. Two others were on joint research council and government committees. One, an academic, was a member of a joint DTI/SERC body which oversaw all the UK strategic research in electronics and compound semiconductor research. The other, a senior researcher from a national laboratory, sat on committees which refereed LINK projects. He identified a difference in review procedures compared to the normal academic process. For example, the advanced semiconductor material LINK programme appraised academic research in one part of the meeting and the collaborative programmes in another. In his opinion, work that was supported on the collaborative projects would not get past the peer review process. It was not that universities put in second-rate proposals. The reason that universities' contribution was limited was that industry was writing the proposals. Firms decided what the work was to do and the timescale for the projects. The problem lay in the fact that LINK projects had to be led by an industrial company. This generally meant that industry wanted universities to undertake short-term research. Most projects, and not by coincidence, were running with the university's contribution at precisely 10 per cent, the minimum required. In the respondent's view, industry needed to be educated to support the notion that something useful can be done with that 10 per cent.

3. Summary and conclusions

The particular characteristic of the interface from the point of view of this sample is the dependency on industry for research funding for materials, industrial expertise, equipment, information and funding. In effect interdependency between industry and the PSSB were regulated by the state; organizations determined the status and function of institutions and those which allocated resources to both and the interface between them. National differences in the composition of the science base, in the instruments of regulation (for example funding programmes, and changes to the status and function of national laboratories) as well as industrial strategies had brought increasing diversity in the ways in which industry and the PSSB worked together in each country. However, the different forms which interaction took were only variations to the main pattern which was that universities and national laboratories were increasingly obliged to undertake short-term projects in order to support longer-term research.

The most radical change in the nature of the interdependence between industry and the PSSB was in the relationship between national laboratories and industry in the UK. Whereas IMEC and LETI were established with the aim of either undertaking research directly relevant to industry or to commercializing existing research, both Harwell and Malvern were required to compete with industry for a variety of sources of income instead of being in complementary research relationships based on their traditional 'public good' role.

The geography of innovation filieres resulting from increasing interdepency between the PSSB and industry in the electronics field was that of more obvious interlocking spatial hierarchies of interaction than were apparent in the flow measurement industry. They intersected because what happened at one level inhibited or facilitated possibilities for individuals and groups in the PSSB to engage with industry at another. At the top were sets of linkages at the European-wide scale sustained by EU programmes (such as ESPRIT and BRITE) and other

European initiatives such as JESSI (a EUREKA programme) which encompassed leading groups from universities and national laboratories forming filiere based on particular technological fields such as software and ASICs. These broader innovation filieres were regulated by the choice of specific targets set at the European level, access criteria and the process of selection and not by objectives relating to protecting the interests of individual firms although that necessarily occurred by default. Most of this sample were participants at this spatial scale, whereas it was shown in Chapter 6 that few individuals/groups in that part of each country's science base were members of networks which extended beyond the national boundary.

The next level down is *innovation* filiere operating within national spaces. Here the main regulators were those which determined which types of funding maintained links within national boundaries. The most important were research council grants, MoD contracts, in the UK national laboratory awards, and projects undertaken on behalf of or in conjunction with industry. At this level there was a marked difference between the UK and France, where there was a high degree of interaction with domestically owned companies (because there were more of them), and Belgium (where the industry base was smaller and largely foreign-owned) where firms based in neighbouring countries and further afield tapped into the strong electronics base.

The third layer was the regional cluster. The fragmented system of university research meant that there was limited evidence of spatial concentration of interaction in the UK. There was more evidence in France, around LETI and in Belgium, focused on IMEC. In France this had grown up as a result of the historical accumulation of expertise in the region mentioned in Chapter 3 and by public funding. In the case of IMEC too the financial factor was of key importance. The priorities set by the Flanders government determined how IMEC's resources would be deployed. This demonstrates how local accountability can operate in tandem with participation in filieres operating at the broadest geographical scales.

Part Four
Conclusions

9
Innovation Filieres and the Geography of Innovation

Introduction

This concluding chapter reviews the empirical evidence presented in Chapters 5 to 8 on national and sectoral differences in the organization of innovation filieres. It compares patterns of technology transfer in the electronic components and flow measurement industries and considers the interdependence between the factors which were responsible for changing patterns of technology transfer and the geographical construction of innovation filieres. The evidence shows that the resulting geographical forms which filieres took in each sector and in each country were quite varied. At the one extreme, innovation filieres were geographically focused on centres of excellence, were strengthened by a plurality of different sources of innovation (see Metcalfe 1993 cited in Cowling and Sugden 1998, 257) and included interactions which incorporated flows of information acquired through networks of innovators in other countries. At the other, filieres were geographically static consisting primarily of externalized interactions with a narrow range of institutions within the national system of innovation. In between were loosely constructed filieres which were spatially diffuse and constructed of an eclectic set of interactions with participants drawing on information from different places at different times as the need arose.

The first section provides an overview of patterns of interaction and changes over time. The second examines similarities and differences in technology transfer processes between the two sectors in each country. The third summarizes the experiences of individual firms and institutions. The fourth reviews the geographical organization of linkages. The last considers the implications of the study.

1.   Patterns of interaction and changes over time

The evidence from the two sector samples in this section is consistent with the summary of innovation patterns in Chapter 4 which laid the foundations for testing the hypothesis that there could be differences in interaction between the two sectors. This identified the key difference between the two sectors as being the relative conformity in the pace of technological change within the flow measurement industry and the relative diversity in the electronic components industry. In the electronic components sector, rapid advances in such areas as materials technology contributed to demand for PSSB expertise to complement in-house R&D enabling firms to compete in developing markets in automobile and space exploration industries. On the basis of the characteristics of the national innovation systems discussed in Chapter 2, which provided the geographical contexts to the hypothesis that there would be differences in patterns of interaction between the three countries, it was anticipated that there would be a low level of links with universities in France in general and in the flow measurement industry in Belgium, and high levels of links with the PSSB base as a whole in the electronic components industry.

1.1   Patterns of interaction

This section begins by summarizing the data on the number of firms in each sector in each country which have links with universities and national laboratories. The overall levels of interaction are recorded in Table 9.1 which includes formal and informal contacts. The findings revealed some sharp contrasts between the flow measurement and electronic components industries and some inter-country similarities:

1   The electronic components sample as a whole had more frequent links with the PSSB than the flow measurement sample.
2   Firms in both sectors were more likely to have links with universities than with national laboratories.
3   Electronic components firms were five times as likely to have links with national laboratories than those in the flow measurement industry.
4   Taking both sectors together, proportionally more UK firms had links with the PSSB than those in France and Belgium.
5   In the flow measurement industry, links with universities were most frequent in the UK and least frequent in Belgium.

6 In the electronic components sector, the level of links with both types of institutions in the UK and Belgian samples was considerably higher than in France.

The percentage of firms in France with links with universities in both sectors was higher than might have been expected from the description of the French innovation system but this is likely to be caused by the small sample size.

Table 9.1 The flow measuring industry and electronic components samples: links with universities and national laboratories

	Flow measurement			Electronic components		
	Sample size	University link number per cent[*]	National laboratory link number per cent[*]	Sample size	University links number per cent[*]	National laboratory link number per cent[*]
UK	13	12 (92)	2 (158)	11	9 (82)	6 (55)
France	5	3 (60)	1 (20)	5	3 (60)	1 (20)
Belgium	7	1 (14)	0 (0)	7	6 (86)	2 (29)
Total	25	16 (64)	2 (8)	23	18 (78)	9 (39)

[1] Percentage of national sample

2. Technology transfer processes

This section discusses sectoral differences in what motivated firms to interact with the science base and the kinds of information they required. It begins by introducing a key element in determining the location of links and the general functioning of filieres, that of how contacts were made.

2.1 Initiation

The study found evidence of two types of industrialists and academics, those who were participants in extensive networks or 'technological communities' spanning industry and the PSSB which were developed and sustained by meeting at conferences and courses; and those which were outside of this kind of professional network.

1 The first type was found in the electronic components sector where high levels of sustained interaction had contributed to a research culture with common sets of expectations about the conduct of externalization and collaboration. The initiative for projects could come from either side. In turn this culture generated positive externalities for firms located in other countries looking to 'plug into' the innovation filiere because a language of co-operation already existed.

2 The second type included engineers and scientists in firms which believed that universities should be more pro-active and expected academics to approach them and academics who thought that firms should be more active in finding out about the expertise in the university system. This group comprised new entrants in the flow measurement industry and branch plant electronic components firms in the UK and in France and some academics in the flow measurement sector. In the flow measurement field, there were information gaps on both sides and less of a research culture for firms and academics to lock into.

2.2 Motives

The evidence from this and the following section on forms of assistance indicates that externalization rather than collaboration was the dominant form of interaction in the flow measurement industry but that this was also one of the range of interactions between the electronic components industry and the PSSB. Table 9.2 summarizes the most common motives for interaction in both sectors. The findings are consistent with other evidence which suggests that there were contrasts between the two sectors in the expectations of what form potential PSSB resources would take, in firms' ability to manage more extensive forms of interaction, and in the roles played by national laboratories as information providers (flow measurement) or as information generators (electronic components). The two main indicators of these contrasts were the extent to which interaction was motivated by an inadequate scale of R&D and the differences in the use of existing information in universities and national laboratories.

First, proportionally more the flow measurement sample than electronic components sample were motivated by inadequate R&D as a factor for working with both universities and national laboratories.

Table 9.2   Most common motives for interaction in the two sectors

	Flow measurement number per cent*		Electronic components number per cent*	
Universities				
Access to new areas of expertise	11	44	12	52
Access to existing information	7	28	10	43
Participation in public programmes	6	24	9	39
Cost saving	6	24	7	30
Inadequate scale of internal R&D	13	52	7	30
Internal problem which could not be solved	7	28	5	22
National labs				
Access to new areas of expertise	7	44	5	22
Access to existing information	7	44	4	17
Participation in public programmes	4	16	4	17
Cost saving	3	12	2	7
Inadequate scale of internal R&D	9	36	4	17
Internal problem which could not be solved	6	24	3	13

* Percentage of total sample
Note: Firms could indicate more than one form of assistance

Secondly, the proportion with contacts with national laboratories for this reason was twice as high in the flow measurement industry than in the electronic components industry. Proportionally more flow measurement firms used existing information from national laboratories than in the electronic components industry but the reverse was true for information from universities. This implies that universities either had a greater stock of information or that information was continually renewed, which meant that firms stayed in touch in order to maintain continuous innovation.

Cost saving was a further factor in motivating interaction. It was more important in interactions with universities than with national laboratories in both sectors and in the UK than in the other countries. It was mentioned by a quarter of flow measurement firms and nearly a third of the electronic components sample. It was a factor only in the UK in the flow measurement industry but was given as a reason by four firms in the UK and by three in the Belgian electronic components samples. This was a much higher level of response than in the Charles and Howells (1992) study. In their sample of 151 UK firms cost saving was the third most important factor behind access to new areas of expertise and inadequate scale of R&D. The significance of cost as a factor can be interpreted as a source of uncertainty within the PSSB as it implies that firms can chose between a variety of options – not to do the work, to do it in-house, or choose between institutions offering similar kinds of expertise.

2.3   Forms of assistance – the industry responses

Table 9.3 shows that the electronic components industry had a higher level of each form of assistance listed not only in the more traditional academic activities such as basic scientific knowledge but also in applied activities such as assistance in new product development, solving development problems and developing testing routines. This

Table 9.3   Most common forms of assistance in the two sectors

	Flow measurement (sample size = 25)		Electronic components (sample size = 23)	
	Number	per cent*	Number	per cent*
Universities				
Basic scientific knowledge	13	52	14	61
New generic technology	6	24	12	52
Solving development problems	6	24	8	35
Assistance in new product development	4	16	9	39
Testing or validations services	3	12	3	13
Developing testing routines	2	8	6	26
Improving product processes	2	8	4	17
National laboratories				
Basic scientific knowledge	1	4	4	17
New generic technology	1	4	5	22
Solving development problems	0	0	3	13
Assistance in new product development	0	0	5	22
Testing or validations services	10	40	2	7
Developing testing routines	5	20	2	7
Improving product processes	1	4	1	4

* Percentage of total sample
Note: Firms could give more than one response

suggests that innovation filieres in the electronic components industry were more integrated than those in the flow measurement industry because of the more extensive range of types of activity linking the articipants.

The most common form of assistance in both sectors was basic scientific knowledge. While more than half the firms with links in each sector gave this response, the proportion was higher in the electronic components industry. On the other hand the sharpest contrast between the two sectors was that the demand for new generic technology was twice as high from the electronic components industry as the flow measurement firms. If the top two categories indicate the use of 'leading edge' research and the bottom five 'technological assistance' in the process of technological change, then the flow measurement industry has required less input into technological advance and product design and development than the electronic components industry. Whereas rapid technological change requires a plurality of interaction including flexible collaboration, incremental change can either be achieved in-house or through externalization of specific tasks up to the moment when markets are lost through the failure to adopt more aggressive innovation strategies.

2.4 Forms of assistance – the PSSB responses

The PSSB samples were asked what was the most common form of assistance which they gave to industry. The highest response from universities in both sectors was basic scientific knowledge. In this respect the PSSB and industry samples were complementary. However, there were country-specific differences within the general pattern. None of the Belgian electronic engineering and applied physics departments said that this was a component of their relationships with industry, whereas all departments in the UK said that it was. Most universities in all three countries and in both sectors provided theoretical input in the form of new generic technology and practical assistance, including specialized testing, solving development problems and providing assistance in product development.

The evidence from the national laboratories was less conclusive. The responses of both samples indicated the prevalence of applied research in their activities. CERT was the only national laboratory which claimed to provide basic scientific knowledge to industry. The dominance of testing and validation activities in the relationships between national laboratories and the flow measurement industry was in sharp contrast with the more technology-based interactions in the electronic components industry. This reflects the priority of conformity to standards as a driver of innovation in the former, compared to competition based on technological lead in the latter. Different levels of uncertainty derive from the basis of the relationship. On the one hand, compliance to standards was then a fairly stable driver of innovation which meant that this was a more predictable source of demand for input from national laboratories while national and EU standards were being established. On the other, demand for theoretical input which is dependent on a range of internal and external forces which shape firms' innovation strategies is potentially less predictable which means that uncertainty within the electronic components industry was externalized into the PSSB.

2.5   Summary

This section has raised issues about processes of change within patterns of externalization and collaboration and how the arrangement of economic uncertainty had shifted with the reorganization of innovation. A primary source of uncertainty arises from the expectations of industry and in the PSSB about what each side wants and the cost that industry was prepared to pay. The responses to questions about what forms of assistance firms sought from the universities and national laboratories and what they provided are indications of confidence and congruence in expectations of what it is that industry and institutions in the PSSB do. The findings suggest some degree of satisfaction in the electronic components firms with the kinds of (industrially relevant) competencies within the PSSB (Gibbons 1992, 99). The major criticism of national laboratories in the UK was not that of technical competence but of the unhelpful attitudes of some personnel. For branch plants in the UK and France which had decided to seek external expertise, the major problem was the lack of information about the kinds of assistance available.

In contrast, there was far less confidence in the flow measurement industry, that the PSSB would be able to provide the right kind of technical support at the right price except where collaborations were co-funded by LINK awards. Some collaborations had failed because of the mis-match of industry and university expectations, with the universities being criticized for failing to take account of industrial priorities. Experiences of interaction had in themselves developed barriers to firms' propensity to network, collaborate and externalize R&D. Some flow measurement firms had learned that it can be difficult and expensive to work with universities and national laboratories because their timescales were much shorter than those of the PSSB. They had learned not to collaborate and not to apply for funds which would subsidize collaboration. Turning Faulkner and Senker's (1995, 231) argument around, that firms only interact with government and academic research when they have specific reasons to do so, there might be good reasons why firms should interact but fail to do so. On the other hand the study showed that there were major educative benefits to new entrants in the UK flow measurement industry which reduced the levels of technological uncertainty in those firms.

Looking at the nature of relationships with industry from the PSSB side, shifts in innovation strategies in both industries in all three countries and the concomitant knock-on effects on the levels of industrial funding increased levels of uncertainty for both universities and national laboratories. The study found that universities were limited in their ability to protect their interests in the face of industrial and political change. Academics were under pressure both internally and externally which take them away from their role as generators of knowledge (David *et al.* 1995). Collaboration and externalization caused information conditions to become less favourable to the PSSB as a result of the changes in the values incorporated into feedback loops within the innovation system. Technology (knowledge) transferred into these institutions from industry changed the expectations and knowledge base of individual researchers, of research groups and of departments as they adapted their activities to meet the needs of industry. This was demonstrated by the finding that in both PSSB samples, scientists and engineers were pragmatic about the necessity of working with industry recognizing the benefits of interaction found in every study of industry and academic links (see for example Faulkner and Senker 1995). These included bringing into universities experience of real world problems which could be used to enhance teaching, provide a source of research ideas and contribute to research funding. More general long-term benefits to industry accrued from ensuring the continuity of the science base as a resource and from the supply of graduates. Moreover, the training element of interaction accrued from co-supervision of students, the sharing of learning through collaboration and industrial residency schemes, and the acquisition of skills required by both sides in the process of arranging short-term projects contributed to the development of groups of people with common sets of technical information and contacts which range from loose occasional associations to the development of elite technological communities.

On the general point about convergence of interest, David and Foray (1994, 13) commented that one of the major lessons which has been learned from studies of success and failure in commercial R&D is that the integration of R&D into business strategies is a crucial condition for sustained performance, and that university-industry collaborations that are not directed to the industry side will be perceived as being ancillary to most of business research expenditures.

The pragmatism expressed by academics in the electronic components industry did not disguise the existence of tensions arising from the commonly held belief that universities were not there to solve the problems of industry. The role of academics was to ask wider and more 'curiosity driven' questions than were of immediate interest to industry. These are issues discussed in other studies including Charles and Howells (1992, 181). Academics in the flow measurement sample also recognized that the developing norm was interaction with industry but were faced with firms unwilling to pay for work which could improve their competitive position. Most firms wanted short-term projects or wanted assistance without paying for the expertise.

3.   Factors influencing firms' decisions to interact over time

This section examines how the forms of engagement between industry and the PSSB are affected by firms' specific knowledge requirements. It does this by discussing the significance of firm level factors, sectoral characteristics, the role of users, the movement of individuals between firms and institutions, and the role of industrial organizations. The main hypothesis was that externalization of innovation is increasing. The evidence is generally consistent with that assumption. There is also evidence that some firms had never worked with institutions in the PSSB and that others were choosing to discontinue doing so.

3.1   Firm factors

This study revealed some patterns common to each sector as well as some major differences between them.

Four distinct groups of firms were found:

A. Those which had institutionalized sets of linkages over a long period of time

These included leading-edge electronic components firms and the research laboratories of flow measurement companies.

B. Those firms which had either recently started to seek external expertise or had increased the scale and scope of interaction from a low base

This is a feature of the flow measurement industry in all three countries and to a lesser extent of branch plants of electronic components firms in the UK and France. In the former it includes new entrants to the field developing new techniques to monitor flows. These were mainly independent firms which had switched to manufacturing flow meters from other forms of instrumentation or needed help with a technological advance.

C. A few firms which had either substantially reduced their level of interaction or had withdrawn completely from interaction

The majority of these were firms in the UK flow measurement industry, both manufacturers and users. Five of the larger UK firms had decreased the level of their direct contact over the previous five years. In each case the enthusiasm of individuals in technical departments for maintaining in-house innovation and links with universities and national laboratories was not matched by the new corporate strategy. Other firms reducing their level of interaction were larger UK electronic components manufacturers who had moved out of silicon research or had refocused their R&D strategies following corporate reorganization.

D. Those which had never worked with universities or national laboratories

Most of these were branch plants of electronic component firms in each country and the sales offices in Belgium of flow meter manufacturers.

Nationality was not a constant factor in whether companies initiated, sustained or decreased links with the science base. Acquisition always affected innovation strategies. Sometimes it resulted in resources allocated to innovation and collaboration increasing and in other cases levels of interaction had been reduced as a consequence of merger. For example, in the flow measurement industry, links with a university on a new sensor had been safeguarded following acquisition which had returned the firm to UK from US ownership. In the electronic components sector acquisition by foreign firms in the UK and in Belgium had strengthened linkages. For example a new Canadian parent of a UK electronics firm had shown a greater commitment and readiness to invest long-term research than was the case under British ownership. In this case acquisition was related to the parent firm's strategic view of how it would compete in the European telecoms market. In another example, the merging of R&D activity in what had been two separate UK companies meant that one set of innovators was displaced when their responsibilities were moved to another part of the company. Take-over was not a factor in decisions of whether firms interacted with the science base in France where patterns of interaction were found to be stable.

Factors which had caused a reduction in links were varied. The re-orientation of research out of silicon in one of the major UK firms in the sample had meant that traditional patterns of links between industry and the science base had been disrupted. The reduction in industrial research in silicon and increased effort in new materials resulted in a re-focusing of that section of the innovation filiere as the PSSB accommodated and responded to changing demands. For one department this meant focusing on theoretical rather than applied research. In Belgium, a small number of firms had reduced the level of interaction. This was because their market share had been eroded by increased competition which in turn had reduced their profitability and money to spend on R&D. The research centre of a telecoms firm had reduced the number of institutions with which it interacted in order to concentrate on linkages a small number of firms.

3.2   Role of users

In the flow measurement industry until the late 1980s the relationship between users and manufacturers had been a core strand of the sector innovation filiere. In recent years this had become a declining component of the innovation/production system. The call for new meter design by British Gas showed clearly that the PSSB was not an institutionalized component of technological change in that section of the flow measurement industry. In the electronic components industry, relationships with individual users were not mentioned as being a prime element in innovation activities or as a decisive factor in whether to interact with universities and/or national laboratories but the demands of users were the major driving force in innovation strategies.

3.3   Mobility

One of the main differences between the sectors was the extent of mobility of scientists and engineers between the PSSB and industry. This was not a factor in the organization of innovation filieres in the flow measurement industry but was a significant albeit changing element in the construction of the filiere in the electronic components industry. In the flow measurement industry mobility between firms and the PSSB was low. In most firms R&D tended to mean incremental engineering developments. This meant that there was less propensity than in R&D intensive industries to recruit from universities and national laboratories. Likewise there was limited potential for reverse flows. Thus possibilities for extending networks and increasing the level of flexible collaboration were restricted.

The major change in the electronic components industry was in the direction of recruitment. In contrast to the traditional pattern of electronic components firms recruiting from universities, in the UK there was a marked trend of senior scientists leaving industry to take university chairs in order to pursue more theoretical research interests. This trend reflected the restructuring of research in major firms, in particular the cessation of research in silicon, which diminished the possibility of industry providing 'rewards and career paths for scientists and engineers which closely parallel those offered by universities' (David *et al.* 1994, 20).

Other forms of mobility were important in creating networks in this sector. The involvement of graduate students on CASE (UK) and CIFRE (France) schemes provided new researchers with experience of both industry and the PSSB. New firm formation strategies by which personnel in IMEC and LETI (but not UK laboratories) were encouraged to form spin-off companies contributed to flows of technology and people out of laboratories.

3.4 Industry groups

A significant difference between the flow measuring industry and the electronic components industry was the institutionalized arrangements through which interaction between industry, universities and national laboratories took place. In the flow measuring industry important examples of institutionalized forms were the Flow Liaison Club and FLOMIC. FLOMIC was an attempt to overcome fragmentation within the industry in Europe. For a time, this was the nexus of interaction on an institutionalized basis. However, some important players in the industry, for example firms in the food industry, chose not to join. Its legacy will be the networks it developed and sustained. Although coalitions were forged temporarily through FLOMIC this organization did not act as a power base to lobby for the industry's interests to be represented in the decision making arena.

There was no equivalent to FLOMIC in the electronic components industry. While leading edge firms may be less in need of a FLOMIC-type structure, there is an argument for its equivalent in the less technologically sophisticated part of the market. Interviews with trade associations and the firms found that they were not organizing agents in either sector of any country. The closest to this model was ISHM which represented manufacturers and academics in a declining section of the industry.

3.5   Summary

The innovation strategies adopted by individual firms and groups of firms combined to reconfigure the structure and geographical organization of innovation filieres. Traditionally some firms and institutions assumed central positions within filieres, others were on the margins. The study shows how over time new categories of firms took on more powerful positions of influence, others withdrew from certain activities to become more peripheral in the structure of innovation filiere while others remained dominant firms. In turn these actions either re-inforced or diverted processes involved in gearing the science base to or from particular priorities, and influenced the extent to which networks were based within the national system of innovation or were more spatially extensive. For example, in flow measurement, privatization of the utilities in the UK provided the opportunity for a water company to participate in the UK and potentially more extensive filieres as an innovator, whereas the withdrawal of user firms from meter design has taken this category of actor to the periphery from a previously central position. In the electronic components industry, the main change was in the strength and coherence of the three national filieres. Recent changes such as defence restructuring in the UK, the increasing opportunities in telecoms and automobile markets, particularly in the UK and Belgium, had meant some re-orientation of some relationships and changes in the composition of the filiere as new firms participated in collaborative activities. Clear trends were that both major companies and smaller firms were strengthening existing relationships but with a smaller number of universities, were placing more emphasis on recruitment, and were spending less on speculative research and were becoming more focused in specific problem solving. As a result of the last trend, interactions were becoming less science-based and more technological in orientation. A particular problem for some UK universities was the withdrawal of firms from research into and production of silicon chips. Although there was considerable expertise in the UK science base in this field this was insufficient to sustain industrial presence. Other firms had entered the filiere as a result of improved profitability for example in France, while others were looking to increase interaction from an existing high level, thus becoming more centrally placed. Centrally placed electronic components firms were in contact with different constituencies including universities and government laboratories, both inside and outside the home country directly through collaboration and through meeting in committees.

Individual operations such as research centres, in both electronic components and flow measurement industries, played important central roles within filieres. Their influence was spatial as well as positional because they were both required to identify the best places for research in different technologies and displaced R&D in branch plants both in the home country and in other countries. On the other hand organizational change, as an example from the UK electronic components sector showed, also meant that universities became intermediary organizations replacing some of the functions of research centres.

Marginal firms were low innovators, maintained low level contact with a small number of institutions and had less developed networks of contacts with other firms as indicated by a lack of involvement in collaborative programmes. As a result of decreased involvement with other players, they had less influence over the pace and direction of technological change in their industry. In some firms, it was the actions of key individuals which sustained interaction with the PSSB and the system as a whole rather than strategic decisions made by the firms.

4.   The geography of innovation: the links between industry and the PSSB

At any moment in time the geography of innovation is comprised of three geographical features – where the firms are, the location of potentially useful existing knowledge and where new knowledge is being generated. Changes in the spatial organization of innovation filiere occur through decisions made by firms, domestic and foreign, on the extent to which they will interact with research institutions within the country in which they are located compared to institutions in other countries. These decisions produce one map of innovation. The pattern of where PSSB scientists and engineers collaborate with industry produces another. Where these systems interlock or exist in parallel is indicative of the degree of coherence in innovation systems in particular countries and sectors. The geographical pattern of national and international linkages of both firms and PSSB institutions is shown in Table 9.4.

Both the industrial and academic communities in the electronic components industry had more geographically diverse links than in the flow measurement industry. Both had more links with organizations within their own country and were more likely to engage in international technology transfer activities. There were some national differences. Electronic components firms in the UK and Belgium were more likely to have links with non-national PSSB institutions than those in the flow measurement industry. On the other hand, Chapters 6 and 8 revealed that there was no consistent pattern when the PSSB samples were compared. For example, the PSSB flow measurement sample were more likely to have international links than the firms but not in France, while UK academics working with electronic components firms had fewer international links than their French and Belgian counterparts. On the other hand, organizational change in the UK electronics industry meant that academics were increasingly looking for foreign industrial partners to replace funding from UK firms. Access to EU funding was a major factor (see below) in the extent of international co-operation. This also reflected pre-existing information about potential partners and the possibilities of access through existing networks.

Table 9.4 Links with PSSB within and external to national innovation system.

| | National research link Number per cent* | | International industrial link Number per cent* | |
	Flow measurement	Electronics	Flow measurement	Electronics
UK				
Firms	17 (68)	18 (78)	1 (8)	6 (55)
Academics	4 (57)	8 (100)	3 (25)	3 (43)
France				
Firms	3 (60)	4 (80)	2 (40)	1 (20)
Academics	3 (100)	2 (100)	0 (0)	2 (100)
Belgium				
Firms	1 (14)	5 (71)	0 (0)	3 (43)
Academics	2 (100)	4 (100)	1 (50)	4 (100)

* Percentage of national sample

The innovation filieres in each sector in France were the least internationally focused. It was mainly that the research centres in both sectors were linked into more geographically extensive innovation networks. In the flow measurement industry, two important organizations, Gaz de France and CERT, participated in both French and UK innovation systems maintaining links with individual universities and more generally through FLOMIC. In the electronic components industry, externalization to LETI was largely confined to firms inside France by government policy which restricted the licensing of technology to non-French firms. In contrast all of the Belgian PSSB electronic components sample were funded directly by foreign firms and three-quarters participated in EU programmes.

The study showed that there was both more clustering of linkages around major centres of expertise and more international activity in the electronic components sector than in the flow measurement industry (cf Breschi and Malerba 1997) but this varied between the countries. There was evidence of clustering in the electronic components sector in Belgium in Flanders, and in Paris and South East France. Local links were not a feature of the UK electronics filiere. Moreover, the openness of the Belgian economy and institutions to foreign companies meant a greater degree of relocation of research activities, particularly to Flanders, than to France or the UK, re-inforcing the importance of the region as a centre of research in microelectronics.

5. Regulation

One of the key themes in the study was the impact of regulation on the structure of innovation filieres. This section discusses how their changing geography is related to regulatory influences. These include spatial policy, funding of innovation, rules and regulations and institutional change.

5.1 Spatial policy

The study focused on two kinds of spatial policy: those sections of the regulatory system which determined both the location and the geographical focus of technology transfer to industry from research in the PSSB, and regionally focused programmes designed to encourage technology transfer. Until very recently, the UK has had no explicit national policy for supporting regionally focused interaction. The fragmented pattern of expertise in a large number of institutions in the UK and the absence of regionally focused innovation support strategies has meant that neither the universities nor national laboratories produced, and in the case of the latter has only recently been intended to provide, a local focus of technological linkages. In contrast, experience in both France and Belgium, particularly with respect to the two major research centres, LETI and IMEC, showed how national and regional initiatives can be effective in building an interface between major research centres and industry by subsidizing the placement of scientists and engineers in complementary environments.

The most successful regionally focused technology transfer initiative was in Belgium. The electronic components firms benefited from schemes organized by the Flanders government to support telecoms and microelectronics and because the regional funding agency, the IWT, had facilitated entry to EU programmes by assistance in the application process. In France, even programmes organized at the regional level by ANVAR and the DRRT had not been used by firms in either industry.

5.2   Funding

The costs of interaction, the level of financing of innovation by industry and by government agencies is a major means by which the quality (content and duration), structure (kinds of participants – firms and institutions) and geography of filieres are regulated. The study showed that innovation filieres were sustained to varying extent by government research and by industry funding. This in turn gave rise to and reflected the relative costs of undertaking research in industry and compared to externalization to the PSSB in different places. The responses from the industry and PSSB samples showed differences in the levels and kinds of funding between the sectors and academic/national laboratory groups (Tables 9.5. and 9.6).

Taking the levels of industry funding first, the general finding was that although over three-quarters of the department in both PSSB

Table 9.5   Funding arrangements in the two industrial samples

	Flow measurement (sample size = 25 firms)		Electronic components (sample size = 23 firms)	
	Number	Per cent*	Number	Per cent*
National funding	6	24	15	65
EC funding	1	4	10	43

* Percentage of total sample

Table 9.6 Funding arrangements in the two PSSB samples

	Flow measurement (sample size = 12 departments)		Electronic components (sample size = 14 departments)	
	Number	Per Cent*	Number	Per cent*
National research institute funding	5	42	9	64
Government industrial programmes	4	33	10	71
EC funding	3	25	8	57
Direct industrial funding	9	75	12	85

* Percentage of total sample

samples had income from industry, this tended to be for much smaller amounts in the flow measurement industry, particularly in the UK, for example through FLOMIC projects, than in the electronic components industry. CERT had negotiated longer-term projects after years of short-term arrangements. However, there was evidence in both sectors of the cost advantage of payment for short-term problem solving activities. Some departments were undertaking short-term projects to service larger projects while for others the sole motivation was income generation. For example, Belgian academic departments in both samples were less certain of obtaining funding from industry for two to three year projects than those in the UK and were obliged to undertake short-term projects. A feature of the Belgian system was the high level of subsidy from the state for industrial research in universities.

Evidence from Chapters 5 and 6 showed that awarding of government or EU grants was of major significance in determining the kinds of interaction firms had with universities and national laboratories. This was particularly the case for flow measurement firms in the UK and Belgium where public support was the key factor in whether a particular development or line of investigation started or the speed at which a project was completed for only a minority of firms. In addition R&D departments of UK firms with LINK awards benefited from the leverage provided by receipt of funds on the decision makers within the firm. There was also the commercial potential derived from the accrued status of working with experts in the PSSB in the field of flow measurement. The evidence also suggested that the system of innovation was likely to suffer from a lack of research renewal. This was because direct industry funding from firms within the home country was the dominant source of research income for each of the PSSB samples which tended to be for short-term projects rather than longer-term research programmes. The agenda for developing new expertise in the PSSB had not been convincingly established at either the national or European scale.

In contrast, the electronic components samples both had appreciably higher levels of funding under national programmes, considerably greater access to EU funds than the flow measurement industry. More UK sample firms had either or both national and EU grants (73 per cent and 55 per cent respectively) than those in Belgium (71 per cent and 43 per cent) and France (40 per cent and 11 per cent). This was associated with the composition of the French sample which contained fewer firms undertaking R&D than in the UK and Belgian samples. Most PSSB departments were undertaking two to three year projects under national awards such as LINK and CASE studentships and EU programmes such as ESPRIT.

The survey showed that access to national and EU funding was a major factor in decisions to commit to longer-term projects. In both sectors, the academic/national laboratory groups had higher levels of participation in both EU and national government programmes than the industrial samples. The difference between the two academic samples was the least marked of all cross-comparisons. The electronic components groups were the most successful in obtaining funds from national research funding bodies and the highest level of EU funding of the four samples. Firms in France, other than the largest and most powerful, did not have access to government funds. In the UK previous innovation support mechanisms such as SFI and ALVEY, whatever their failings, were generally seen as being far more appropriate to industry's needs than current programmes. LINK was criticized by firms and academics in both sectors and not just by those which had failed to get funding.

The difference in access to external funding from the EU was an important mediating factor in the extent to which domestic innovation filieres were linked into more geographical extensive production/ innovation systems. The low levels of international linkage suggests that flow measurement innovation systems were less efficient than those in the electronic components industry. This assumption is based on the evidence that international links are the means by which firms acquire know-how which is lacking at home (Archibugi and Michie 1995). Those innovation filiere which do not relocate information from other countries into the domestic system are therefore less efficient because advances in technology made in other countries are not incorporated into in-house activities through flexible collaboration.

In sum, the examples from the study illustrated how funding of collaborative research projects between industry and universities and national laboratories, as well as the general levels of funding for these institutions, constitute politically constructed terms of trade. Funding priorities are indicative of the power of particular groups to attract funds and/or of their significance within national and international innovations systems. Firms in the electronic components industry had a greater capacity to influence priorities in the funding systems than firms in the flow measurement industry through their greater purchasing power. This was in part to do with the lower propensity on the part of the latter industry to articulate demand for resources as well as the size of the sector. In the electronic components industry, networks were extended by international collaboration, in part because of decisions made on applications to EU programmes.

The UK government's lack of support for mature industries such as the flow measuring industry and their low priority within the EU have a strategic importance far greater than their own immediate market has significant knock-on effects for other sectors. Inadequate funding contributes to technology lock-in where possibilities of supporting new developments are missed which would enhance profitability in both producers and users. Moreover, the problem for the flow measurement industry was that radical change in the development of alternative solutions to metering including the replacement of meters by in-pipe sensors requires a new skill base. This needed greater cross-disciplinary interaction but was not being funded either by industry or by the government.

5.3 Representation of interests

The degree of leverage exercised by industry over access to research and to individuals in the science base is in part determined by such institutional arrangements as the membership of grant awarding committees which determine the content of research programmes and which get funded. Membership of committees is an important indicator of which firms and/or individuals are identified within the system as being important and of whose interests are being represented. This is also dependent on the willingness of individuals/firms to become part of the system. While firms in the UK electronic components industry identified membership as being an important means of monitoring and influencing developments in technology, industrialists in the flow measurement industry did not become involved with this kind of committee and were more commonly represented on standards committees. This meant that the latter industry was less well represented in networks of power and influence in which the direction of research in the PSSB was decided. There was no reported equivalent set of committee structures in either France or Belgium. The closest case was where an academic from one of the Belgian universities was on the board of IMEC.

5.4   Institutional change

Changes in institutional responsibilities were of major significance in influencing the structure and geographical focus of innovation filieres. While universities were by far the main link in innovation filieres in each sector in each country, national laboratories were an integral part of the innovation filieres in electronic components technologies in each country and in the flow measurement industry in the UK and France.

The central position played by national/regional laboratories in electronics research in the UK, indicated in Chapter 3, is reflected in the finding in Chapter 8 that over half of the sample collaborated with either or both Malvern or Harwell laboratories. The lower figures for Belgium and France reflect the size and composition of the samples as they are inconsistent with information provided by respondents from IMEC and LETI. The central position formally occupied by NEL in the UK flow measurement filiere was declining. The small number of firms studied which collaborated with NEL was associated with the laboratory's decreasing research role and firms' dissatisfaction with previous experiences of collaboration.

In both sectors, the principle of increased 'efficiency' which lay behind the moves to prioritize commercial objectives over long-term research in UK and French national laboratories had changed the relationship between them and the universities from a co-operative to a competitive model. Laboratories such as NEL and CERT in flow measurement and Malvern and LETI in electronics had become increasingly in competition with universities for the provision of specialist expertise to industry. In practice firms found that laboratories were still more expensive than universities. More generally, as indicated by the respondent from the UK water company, universities and national laboratories were increasingly in competition with similar organizations in other countries as firms widened the scope of their search for technology.

Another dimension of institutional change was the increasing commodification of university and national laboratory research through the rules on ownership of IP by individual organisations as well as by funding bodies. This was an important aspect of the way innovation filieres functioned. It was the norm for electronic components firms to own the IP generated in collaborations with universities. This appeared to be accepted as the status quo by the academic sample. In some cases this was because it represented a trade-off of expertise for research income and in others because the academics believed the research they were doing was of no immediate commercial value. However, one of the MNCs in the UK sample thought that this situation was undesirable. In the flow measurement industry, academics were less willing to release control of their IP. This was mentioned as an issue by over half the UK firms. One reason for this was that the financial rewards for the work undertaken by industry were so limited and control of IP meant that there might be opportunities for undertaking further industrial research in that area.

The restrictions imposed by industry on publications was more of an issue for the academics in the UK than ownership of IP. Chapter 8 gave examples of how failure to publish had caused serious problems in the UK for both individual academics and the Microfabrication Facility.

6. Conclusions and implications

The approach in this book has been to examine innovation as a component of production systems, focusing on the way links between industry and the PSSB are determined by firms' responses to the challenges of technological change. Preston's (1989) definition of a filiere provided the basis of analysis. This required that a filiere be seen as a 'structure in process', being subject to changes in technology, market structure or government regulation. This book has built on this concept by showing that innovation filieres are constructed from interactions between industry and universities and national laboratories consisting of networks, flexible collaboration and externalization. The usefulness of the filiere concept to understanding the spatial dynamics of the innovation process is that it is based on the notion that the interdependence between a combination of factors such as technological change and regulation are implicated in firms' responses to their external competitive environment. The idea of interdependence can be extended to identification of how the location of activities and the linkages between them are affected by the accumulated specific techno-industrial features of places and countries as well as more the more spatially extensive pre-existing geographical construction of sectors.

The study has a number of implications for policy makers. The book illustrates what happens when policy makers do not support interaction and the benefits when they do. Public policy has produced different levels of stability and uncertainty within innovation systems in relation to both industry and the science base. Stability has been achieved by support for long-term research and long-term funding programmes. At its most positive stability allows selected actors in 'elite coalitions' to lobby on the behalf of particular sectors through access to centres of influence. An example of this is the case of the electronic components industry in Belgium which has benefited from the long-term commitment to IMEC by the Flanders government. This form of spatial policy is of particular interest as a means of altering the locus of technological change. IMEC is more than a research centre, it is a political symbol and represents powerful industrial interests. The laboratory's commitment to long-term research and its international collaborations has meant that the organization is providing locally generated research and acting as a technological antenna picking up signals from the outside world. The benefits of these activities have been distributed locally through its various innovation support activities. However, even IMEC was facing pressures to become more short-term orientated.

This comment leads to a general concern expressed by firms and academics interviewed in both industries which was the threat to longer-term research in both universities and national laboratories. In the UK, an additional dimension to this problem resulted from the redefinition of the roles universities and national laboratories play. Changing their status and the realignment of respective roles *vis-à-vis* industry was introducing greater competition to the PSSB sector, thereby increasing uncertainty in academia about their ability to attract research funding, particularly in departments where equipment was not being updated.

The increasing interdependence between industry and PSSB as a result of the development of innovation filieres has a number of implications for the direction of scientific research. The longer-term involvement from the electronic components sector compared to that of the flow measuring industry may mean that the former industry's influence over the direction of research in universities is more apparent than in the flow measuring sector. Moreover, the increasing industrial influence over academic research can be a wasteful process. Selective industrial funding restricts the exploitation of ideas, since they must go undeveloped if there is no money to further them. Resources in people and facilities are not being used and creativity is being stifled. This constitutes an unresolved problem in all three countries.

References

ACARD (1983) *Improving Research Links Between Higher Education and Industry*, London: HMSO

ACOST (1991) *Science and Technology Issues: a Review by ACOST* London: ACOST

Adler, P. (1990) 'Shared Learning' *Management Science* 36 (8) pp. 938–57

Albertini, S. and Butler, J. (1997) 'The Types of Knowledge Used in R&D Networking and Innovation Activities' Ch 1 in J. Butler and A. Piccaluga (eds) *Knowledge, Technology and Innovative Organisations* Milano: Guerni E. Associati

Allen, T. J. (1977) *Managing the Flow of Technology*, Cambridge, MA: MIT Press

Amin A. and Thrift N. (Eds) (1994) *Globalization, Institutions and Regional Development in Europe* Oxford: University Press

Amin, A. and Thrift, N. (1995) 'Globalisation, Institutional "Thickness" and the Local Economy' Ch 4 in P. Healey, S. Cameron, S. Davoudi, S. Graham and A. Madani-Pour (eds) *Managing Cities: the New Urban Context* Chichester: John Wiley and Sons pp. 91–108

Angell, C. A. Collins G. C. S. Jones, A. D. W. and Quinn, J. J. (1985*) Information Transfer in Engineering and Science* London: Technical Change Centre

Anselin, L., Varga, A. and Acs, Z. (1997) 'Entrepreneurship, Geographic Spillovers and University Research: a Spatial Economic Approach' Paper presented at CBR workshop, Cambridge UK, March 1997

Archibugi, D. and Pianta, M. (1992) *The Technological Specialisation of Advanced Countries, A report on the EEC International Science and Technology Activities* Dordrecht: Kluwer Academic Publishers

Archibugi, D. and Michie, J. (1995) 'The Globalisation of Technology: a New Taxonomy' *Cambridge Journal of Economics* 19, 1 February 1995 pp. 121–40

Atkinson, H. Rogers, P. and Bond, R. (1990) *Research in the United Kingdom, France and Germany* Volume 1 Swindon: SERC

Bainbridge, T. and Teasdale, A. (1995) *The Penguin Companion to European Union* London: Penguin

Beckouche, P. (1991) 'French High Tech and Space: a Double Cleavage' Ch 11 in G. Benko and M. Dunford *Industrial Change and Regional Development: the Transformation of New Industrial Spaces* London: Belhaven Press pp. 205–25

Begg, I. (1993) 'Industrial Policy, High technology Industry and the Regions' Ch 12 in R. T. Harrison and M. Hart (eds) *Spatial Policy in a Divided Nation* London: Jessica Kingsley pp. 216–31

Bergman, E. M. Maier, G. and F. Todtling (eds) (1991) *Regions Reconsidered* London: Mansell Publishing

Bonaccorsi, A. and Piccaluga, A. (1994) 'A Theoretical Framework for the Evaluation of University-Industry Relationships' *R&D Management* 24, (3) pp. 229–47

Borde, J. (1992) CNRS, London, personal communication

Bradach, J. L. and Eccles, R. G. 'Price, Authority and Trust: from Ideal Types to Plural Forms' Ch 23 in G. Thompson, J. Frances, R. Levacic, and J. Mitchell (eds) *Markets, Hierarchies & Networks: The Coordination of Social Life* Published in association with the Open University London: Sage Publications pp. 277–92

Branscomb, L. M. (1993) 'National Laboratories: The search for new missions and structures' Ch 4 in L. M. Branscomb (ed) *Empowering Technology: Implementing a U.S. Strategy* Cambridge Mass: The MIT Press pp. 103–34

Breschi, S. and Malerba, F. (1997) 'Sectoral Innovation Systems: Technological Regimes, Schumpeterian Dynamics and Spatial Boundaries' Ch 6 in C. Edquist (ed) *Systems of Innovation: Technologies, Institutions and Organisations* Pinter: London pp. 130–56

Brookes, A.-M. (1994a) 'A Survey of European National Measurement Systems' London: DTI National Measurement System Policy Unit

Brookes, A.-M. (1994b) 'An investigation into the influence of the European environment on a strategy for the UK national measurement system' London: DTI National Measurement System Policy Unit

Brunat, E. and Reverdy, B. (1989) 'Market Forces: Linking University and Industrial Research in France' *Science and Public Policy* October 1989, pp. 283–93

Buswell, R. J. Easterbrook, R. P. and Morpet C. S. (1985) 'Geography, Regions and Research and Development Activity: the Case of the United Kingdom' in A. T. Thwaites and Oakey R. P. (eds) *The Regional Impact of Technological Change* London: Frances Pinter pp. 36–66

Bye, P. and Chanaron J.-J. (1995) 'Technological trajectories and strategies' *International Journal of Technology Management* 10 (1) pp. 45–66

Camagni R. (1990) 'Local Milieu, Uncertainty and Innovation Networks: Towards a New Dynamic Theory of Economic Space', Paper presented at the Seminar on 'Network of Innovators' University of Quebec, Montreal, May 1990 p. 24

Camagni, R. (1994) 'Inter-firm Industrial Networks: the Costs and Benefits of Cooperative Behaviour' *Journal of Industry Studies* 1 (1) pp. 1–15

Casson, M. (ed) (1991) *Global Research Strategy and International Competitiveness* Oxford: Basil Blackwell

Catin M. (1985) *Effets Externes, March, et Systémes de Décision Collective*, Paris: Cujas

Chanaron, J. J. (1989) 'French Science Policy and Local High Tech Industries' *Science and Public Policy* February 1989 pp. 19–26

Charles, D. and Howells, J. (1992) *Technology Transfer in Europe: Public and Private Networks* London: Belhaven Press

Cheshire, P. C., D'Arcy, and Giussani, B. (1992) 'Local, Regional and National Government in Britain: a Dreadful Warning' Paper presented at thirty Second European Congress, Regional Science Association Louvain-La-Neuve, Belgium, 25–8 August 1992

Chesnais, F. (1993) 'The French National System of Innovation' Ch 6 in R. Nelson (ed) *National Innovation Systems: a Comparative Analysis* New York, Oxford: OUP

Clark, G. Gertler, M. and Whiteman, J. (1986) *Regional Dynamics: Studies in Adjustment Theory* Boston: Allen and Unwin

Clark, G. (1992) ''Real' Regulation: the Administrative State' *Environment and Planning A* 24, pp. 615–27

Cohen, W. M. and Levinthal, D. A. (1990) 'Absorptive Capacity: a New Perspective on Learning and Innovation' *Administrative Science Quarterly*, 35 pp. 128–52

Counseil National De La Politique Scientifique (1984) 'Politique de Recherché en Micro-electronique et ses Consequences Sociales' Brussels: Counseil National De La Politique Scientifique

Cooke, P. (1993) 'The New Wave of Regional Innovation Networks: Analysis, Characteristics and Strategy' Mimeo, Regional Industrial Research, Dept. of City & Regional Planning, University of Wales, Cardiff, February 1993

Cooke, P. (1998) 'Global Clustering and Regional Innovation: Systemic Integration in Wales' Ch 10 in H.-J. Braczyk, P. Cooke and M. Heidenreich (eds) *Regional Innovation Systems* London: UCL Press

Cooke P. and Morgan K. (1998) *The Associational Economy* Oxford: Oxford University Press

Cooke, P., Moulaert, F. Swyngedouw, E., Weinstein, O. and Wells, P. (eds) (1992) *Towards Global Localisation* London: UCL Press

Cookson, C. (1990) 'Victorian meter meets its electronic match' *Financial Times* September 4 1990

Cowling, K. and Sugden, R. (1998) 'Technology Policy: New Directions' Ch 12 in J. Michie and J. Grieve Smith (eds) *Globalisation, Growth, and Governance: Creating an Innovative Economy* Oxford: OUP pp. 239–62

David, P. and Foray, D. (1994) 'Accessing and Expanding the Science and Technology Knowledge Base' Working Group on Innovation and Technology Policy, Paris: OECD

David, P., Guena, A. and Steinmeuller, W. E. (1995) 'Additionality as a Principle of European R&D Funding' Maastricht: Merit Working Paper 2/95–012

David, P. Mowery, D. C. and Steinmeuller, W. E. (1994) 'University-Industry Research Collaborations: Managing Missions in Conflict' Paper presented at the CEPR/AAAS conference 'University goals, Institutional Mechanisms, and the 'Industrial Transferability' of Research', sponsored by the American Academy of Arts and Sciences and the Centre for Economic Policy Research at Stanford University, held at Stanford, CA 18–20 March 1994

De Backere, K. and Rappa, M. (1994) 'Technological Communities and the Diffusion of Knowledge: a Replication and validation' *R&D Management* 24, 4 pp. 355–71

De Bernady, M. (1998) 'RTD SMEs and Collective Learning: Historicity and Ability for a Local Economy to Evolve: the Grenoble Case Study' in Regional Reports of the TSER European Network on 'Networks, collective learning and RTD in regionally-clustered high-technology small and medium-sized enterprises' ESRC Centre for Business Research, University of Cambridge, Department of Applied Economics, Sidgwick Avenue Cambridge CB3 9DE

De Bresson, C. and Amesse, F. (1991) 'Networks of Innovators: a Review and Introduction to the Issue' *Research Policy* 20 pp. 363–97

Department of Trade and Industry (1988) *DTI – The Department for Enterprise* London: HMSO

Department of Trade and Industry and PA Technology (1989) *Manufacturing into the 1990's* London: DTI

Department of Trade and Industry (1990) *Country Profile: Belgium* British Overseas Trade Board, London: DTI

Department of Trade and Industry/BIS Mackintosh (1990a) *Electronic Components: a Decade of Change: the European Market to 1995* London: DTI

Department of Trade and Industry/Ernst and Young (1990b) *Electronic Components: a Decade of Change: an Overview of European Users* DTI London: DTI

Department of Trade and Industry (1991) *Country Profile: France* British Overseas Trade Board, London: DTI

Department of Trade and Industry /Confederation of British Industry (1992) 'Innovation: the Best Practice' *DTI Innovation Unit*, London: DTI

Dicken, P. and Thrift, N. (1992) 'The Organisation of Production: Why Business Enterprises Matter in the Study of Geographical Industrialisation' *Transactions* 17 (3) 279–91

Dicken, P. Forsgren M. and Malmberg, P. (1994) 'The Local Embeddedness of Transnational Corporations' Ch 1 in A. Amin and N. Thrift (eds) *Globalisation, Institutions and Regional Development in Europe'* Oxford: OUP pp. 23–45

Dickson, K. (1982) 'Industrial Infrastructure, Innovation and Equipment Supply: the Case of the UK Semiconductor Industry', in *Proceedings of 9th Annual Conference of the European Association for Research into Industrial Economics*, Leuven, Belgium..

Dosi, G. (1984) *Technical Change and Industrial Transformation* Macmillan, London

Dosi, G. and Orsenigo, L. (1988) 'Coordination and Transformation: an Overview of Structures, Behaviours and Change in Evolutionary Environments' Ch 3 in G. Dosi, C. Freeman, R. Nelson, G. Silverberg, and L. Soete (eds) *Technical Change and Economic Theory* London: Pinter pp. 13–37

Dunford, M. (1989) 'Technopoles, Politics and Markets: the Development of Electronics in Grenoble and Silicon Glen' Ch 4 in M. Sharp and P. Holmes (eds) *Strategies for New Technologies: Case Studies from Britain and France* New York, London: Philip Allan pp. 80–118

EC (1994) *The European Report on Science and Technology Indicators* EUR 15897, Luxembourg: EC

Eddison, I. (1993) 'Microelectronics and Optoelectronics' in Joint Framework for Information Technology (JFIT) *Annual Report 1993* DTI/SERC Technology Programmes and Services Division, London: Department of Trade and Industry

Edquist, C. (1997) 'Preface', *Systems of Innovation* London: Pinter

Efficiency Unit (1988) 'Improving Management in Government: the Next Steps' (Ibbs) London: HMSO

Ellison, G. and Glaeser, E. L. (1997) 'Geographic Concentration in U.S. Manufacturing Industries: a Dartboard Approach' *Journal of Political Economy* 105 (5) pp. 889–927

Ergas, H. (1993) 'Europe's Policy for High technology: Has Anything Been Learnt?' Mimeo Paris:OECD

Escorsa, P., Maspons, R. and Valls, J. (1998) 'Regional Report: Catalonia' in Regional Reports of the TSER European Network on 'Networks, collective learning and RTD in regionally-clustered high-technology small and medium-sized enterprises' ESRC Centre for Business Research, University of Cambridge, Department of Applied Economics

Ettlinger, N. (1994) 'The Localisation of Development in Comparative Perspective' *Economic Geography* pp. 66, 67–82

Etzkowitz, H. and Stevens, A. J. (1995) 'Inching towards Industrial Policy: the Universities' Role in Government Initiatives to Assist Small, Innovative Companies in the US *Science Studies*, 8 (2) pp. 13–31

Eureka INTO News December (1994) 'Aims of Eureka' London: DTI Electronics and Engineering Division

Evans, R. Butler, N. and Goncalves, E. (1991) *The Campus Connection: Military Research on Campus* London: CND

Faulkner, W. and Senker, J. (1995) *Knowledge Frontiers: Public Sector Research and Industrial Innovation in Biotechnology, Engineering Ceramics, and Parallel Computing* Oxford: Clarendon Press

Feller, I. (1990) 'Universities as Engines of R&D Based Economic Growth: They Think They Can' *Research Policy*, Vol 19 pp. 335–48

Findlay, A. and Gould, W. T. S. (1989) 'Skilled International Migration: a Research Agenda' *Area* (1989) 21 1, 3–11

FLOMIC (1992) *FLOMIC Mission: Statement and Objectives* Internal document, FLOMIC: Cranfield University Department of Instrumentation and Engineering

Furness, R. A. and Heritage, J. E. (1989) *Redwood Guide to Flowmeters* London: IBC Technical Publications

Fusfeld, L. and Peters, H. (1983) (eds) 'Current U S University-Industry Research Connections' in *University-Industry Research Relationships: Selected Studies* National Science Foundation: Washington DC pp. 1–162

Henderson, E. Ince, M. and MacGregor, K. (1990) 'UK Research Cash Lags Behind' *Times Higher Education Supplement* (THES) September 28 1990 Number 934 p. 1

Garnsey, E. and Lawton Smith, H. (1997) 'The Science-Based Industrial Complex: Diverse Paths: Common Processes' *Research Papers in Management Studies*, WP 12/97 Judge Institute of Management Studies, University of Cambridge

Georghiou, L. Stein, J. A. Janes, M. Senker, J. Pifer, M. Cameron, H. Nedeva, M. Yates, J. and Boden, M. (1992) *The Impact of European Community Policies for Research and Technological Development upon Science and Technology in the United Kingdom* Report prepared for DGXII of the Commission of the European Communities and the United Kingdom Office of Science and Technology, Brussels: EC

Gertler, M. (1992) 'Flexibility Revisited: Districts, Nation States, and Forces of Organisation of Production' *Transactions* 17 (3) pp. 259–78

Gertler, M. (1997) 'The Invention of Regional Culture' in R. Lee and J. Wills (eds) *Geographies of Economics* London: Arnold pp. 47–58

Getimis P. Kafkalas G. (1989) 'Spatial Processes and Forms of Regulation: Locality and Beyond', paper presented at the International Symposium on 'Regulation, Innovation and Spatial Development' 13–15 September, Cardiff

Gibbons, M. (1992) 'The Industrial-Academic Research Agenda' Ch 6 in T. Whiston and R. L. Geiger (eds) *Research and Higher Education: the United Kingdom and the United States* Buckingham: SRHE and Open University Press pp. 89–102

Goddard, J. Thwaites, A. and Gibbs, D. C. (1986) 'The Regional Dimension to Technological Change in Great Britain' Ch 7, in A. Amin and J. B. Goddard (eds) *Technological Change, Industrial Restructuring and Regional Development* London: Allen and Unwin

Gordon R. (1989) 'Alleanze Strategiche e Socializzazione del Capitale' *Economia e Politica Industriale* No. 59, pp. 267–70.

Hall, P. (1985) 'The Geography of the Fifth Kondratieff Cycle' in P. Hall and A. Markusen (eds) *Silicon Landscapes* London: Allen and Unwin pp. 1–19

Hancher, L. and Moran, M. (1989) *Capitalism, Culture, and Economic Regulation* Oxford: Clarendon Press

Harrison, B. (1997) *Lean and Mean* New York: Guilford

Harvey, D. (1982) *Limits to Capital* Oxford: Blackwell

Hay, D. A. and Morris, D. J. (1991) *Industrial Economics and Organisation* Oxford: Oxford University Press

Hayward, A. T. (1991) 'Taking the Measure of Flow' *Professional Engineering* March 1991

Heim, C. E. (1988) 'Government Research Establishments, State Capacity, and Distribution of Industry Policy in Britain' *Regional Studies* 22 pp. 375–86

Higham, E. H. (1987) *An Assessment of the Priorities for Research and Development in Process Instrumentation and Process Control* The Institution of Chemical Engineers, Rugby

Hilpert U. (1991) 'Regional Policy in the Process of Industrial Modernization: the Decentralisation of Innovation by the Regionalization of High Tech' Ch 1 in U. Hilpert (ed) *Regional Innovation and Decentralization: High Tech Industry and Government Policy* London: Routledge pp. 3–34

Hilpert, U. and Ruffieux, B. (1991) 'Innovation, Politics and Regional Development: Technology Parks and Regional Participation in High Tech in France and West Germany' Ch 3 in U. Hilpert (ed.) *Regional Innovation and Decentralisation: High Tech Industry and Government Policy* London: Routledge pp. 61–88

Hippel, E. Von (1978) 'Users as Innovators' *Technology Review* 80 (3), pp. 30–4.

HMSO (1993) *Realising Our Potential* London: HMSO

HMSO (1994) *Competitiveness: Helping Business to Win* London: HMSO

Hobday, M. G. (1991) 'The European Semiconductor Industry: Resurgence and Rationalisation' Ch 5 in C. Freeman, M. Sharp and W. Walker (eds) *Technology and the Future of Europe: Global Competition and the Environment in the 1990s* London: Pinter Publishers pp. 80–94

Howells, J. R. L. (1986) 'Industry-Academic Links in Research and Innovation: a National and Regional Development Perspective' *Regional Studies*, 20 pp. 472–6

Howells, (1990) 'The Internationalisation of R&D and Development of Global Research Networks' *Regional Studies* 24 (6) pp. 495–512

Howells, J. and Green, A. E. (1986) 'Location, Technology and Industrial Organisation in UK Services' *Progress in Planning* 27 pp. 83–184

Hudson, R. (1995) 'Regional Futures: Industrial Restructuring, New Production Concepts and Spatial Development Strategies in the New Europe' Paper presented to the Regional Studies Association European Conference on Regional Futures, Gothenberg, 6–9 May 1995

JFIT News (1993) No. 43 May 1993 London: DTI/SERC Technology Programmes and Services Division, Department of Trade and Industry

Johnston, J. (1993) *New Directions in Flow Measurement* FLOMIC: Cranfield University

Keeble, D. Lawson, C. Lawton Smith, H. Moore, B. and Wilkinson, F. (1997) 'Internationalisation Processes and Networking and Local Embeddedness in Technology-Intensive Small Firms' ESRC Centre for Business Research, Cambridge University WP 53 March 1997

Kenney, M. (1986) *Biotechnology: the University–Industry Complex* New Haven: Yale University Press

Kleinknecht, A. (1987) *Innovation Patterns in Crisis and Prosperity* New York: St. Martin's Press

Läpple, D. (1976) *Staat en Algemene Produktievoorwaarden. Grondslagen voor een Kritiek op de Infrastruktuurtheorieën*, Zone special 1, Stichting Zone, Amsterdam

Lawton Smith, H. (1991) 'Industry and Academic Links: the Case of Oxford University' *Environment and Planning C: Government and Policy* 9 pp. 403–16

Lawton Smith, H. (1991) 'The Role of Incubators in Local Industrial Development: The Cryogenics Industry in Oxfordshire' *Entrepreneurship and Regional Development* 3 (2) April–June, 175–94

Lawton Smith, H. (1993) 'Externalisation of Research and Development in Europe' *European Planning Studies*, Vol 1. No. 4 pp. 465–82

Lawton Smith, H. Dickson, K. and Smith, S. (1991) '"There Are Two Sides to Every Story": Innovation and Collaboration Within Networks of Large and Small Firms' *Research Policy* 20, pp. 457–68

Lawton Smith, H. (1995) 'The Contribution of National Laboratories to the European scientific labour market' *Industry and Higher Education* 9 (3) June 1995 pp. 176–85

Lawton Smith, H. (1997) 'Regulatory Change and Skill Transfer: the Case of National Laboratories in the UK, France and Belgium' *Regional Studies* 31 (1) pp. 41–54

Lawton Smith, H. (1998) 'Barriers to Technology Transfer: Local Impediments in Oxfordshire' *Environment and Planning C: Government and Policy* 16 pp. 433–48

Lawton Smith, H. and De Bernady, M. (1998) 'University and Public Research Institute Links with Regional High-Tech SMEs' Paper presented at *TSER European Network on 'Networks, Collective Learning and RTD in Regionally-clustered High-Technology Small and Medium-sized Enterprises'* Conference, Robinson College, University of Cambridge, 7 December 1998

Lindholm Dahlstrand, A. (1998) 'The Development of Technology-Based SMEs in the Goteborg Region' in *Regional Reports of the TSER European Network on "Networks, Collective Learning and RTD in Regionally-clustered High-Technology Small and Medium-sized Enterprises* Cambridge: ESRC Centre for Business Research, University of Cambridge, Department of Applied Economics

Link, A. N. and Rees, J. (1990) 'Firm Size, University Based Research and the Returns to R&D' *Small Business Economics*, 2 pp. 25–31

LINK Newsletter (1990) 1 August London: DTI

Lipietz, A. (1993) 'The Local and the Global: Regional Individuality or Inter-Regionalism' *Transactions* 18, 1 pp. 8–18

Lundvall, B.-A. (1988) 'Innovation as an Interactive Process – from User-Producer Interaction to National System of Innovation' in G. Dosi, C. Freeman, R. Nelson, G. Silverberg, and L. Soete (eds) *Technical Change and Economic Theory* London: Pinter

Lundvall, B.-A. (1992) (ed) *National Systems of Innovation: Towards a Theory of Innovation and Interactive Learning* London: Pinter

Macdonald, S. and Williams, C. (1994) 'The Survival of the Gatekeeper' *Research Policy* 23, 2 pp. 123–32

Macdonald, S. (1997) 'Formal Collaboration and Informal Information Networks', a note to the SPSG meeting, 24/25 February 1997, London.

Maillat, D. (1991) 'The Innovation Process and the Role of the Milieu' Ch 6 in E. M. Bergman, G. Maier, and F., Todtling (eds) *Regions Reconsidered* London: Mansell Publishing, pp. 103–17

Malecki, E. (1990) 'R&D and Technology Transfer in Economic Development: the Role of Regional Technological Capability' in *The Spatial Context of Technological Development* Aldershot: Avebury pp. 303–30

Malecki, E. (1991) *Technology and Economic Development: the Dynamics of Local, Regional and National Change* Harlow: Longman

Manners, D. (1994) 'The Chips are Down' the *Guardian* December 15 On Line Section page 5.

Marx, K. (1977) *Capital* London: Lawrence and Wishart

Maskell, P. and Malmberg, A. (1995) 'Localised Learning and Industrial Competitiveness' Paper presented at the Regional Studies Association Conference on 'Regional Futures' Gothenburg, 6–9 May 1995

Mason, G. and Wagner, K. (1994) 'Innovation and the Skill Mix: Chemicals and Engineering in Britain and Germany' *National Institute Economic Review*, NIESR May 1994 No. 148 Metcalfe, S. (1998) 'Endogenous Growth and the Innovation Process: Why Institutions Matter' Paper presented at RGS-IBG economic Geography Research Group Seminar on 'Institutions and Governance' 3 July 1998, Department of Geography UCL, London *Ministere De La Recherché et de la Technologie* (1990) Publicity Brochure Paris: MRT MMC(1992) *United Kingdom Atomic Energy Authority: a Report on the Service Provided by the Authority* London: HMSO

Mommen, A. (1994) *The Belgian Economy in the Twentieth Century* London: Routledge

Monopolies and Mergers Commission (1989) *The General Electric Company plc, Siemens AG and the Plessey Company plc: a Report on the Proposed Mergers* London: HMSO

Moore, B. (1996) 'Sources of Innovation, Technology Transfer and Diffusion' Ch 7 *in The Changing State of British Enterprise: Growth, Innovation and Competitive Advantage in Small and Medium Sized Firms 1986–95* ESRC Centre for Business Research, Department for Applied Economics, University of Cambridge, September 1996

Moulaert, F. and Swyngedouw, E. (1992) 'Accumulation and Organisation in Computing and Communications Industries: a Regulationist Approach' Ch 3 in P. Cooke, F. Moulaert, E. Swyngedouw, O. Weinstein, and P. Wells (eds) (1992) *Towards Global Localisation* London: UCL Press pp. 39–60

Moulaert, F. and Willekens, F. (1987) 'Decentralisation Industrial Policy in Belgium: Towards a New Economic Feudalism' in H. Muegger, W. Stohr, P. Hisp and B. Stickey (eds) *International Economic Restructuring and the Regional Community* Aldershot: Avebury pp. 314–36

National Economic Development Council (NEDC) (1988) *Performance and Competitive Success: Strengthening Competitiveness in UK Electronics* London: NEDC

National Economic Development Office (NEDO) (1990) *Electronic Components Sector Group Collaborative Research and Development in Electronics* Project Group Report CR 21 London: NEDO

National Economic Development Council (NEDC) (1991) 'Electronics: Strengthening the United Kingdom's Technological Base' Memorandum by Mr Eric Hammond, OBE, Chairman of NEDC Electronics Component Sector Group NEDC (91)12, 18 March 1991, London: NEDC

National Engineering Laboratory (1995) 'Multiphase Flow Club Workshop' *Flow Tidings* No. 9 April East Kilbride: NEL

National Metering Trials Working Group (1993) *Final Report – Summary* London: Water Services Association, Water Companies Association, Office of Water Services, WRc and the Department of the Environment, Multi-Departmental Scrutiny of Public Sector Research Establishments (1994) London: HMSO

Nelson, R. (1986) *High Technology Policies: a Five-Nation Comparison* Studies in Economic Policy, Washington DC and London: American Enterprise Institute,

Nelson, R. (1989) 'What is Private and What is Public about Technology?' *Science, Technology and Human Values*, 14 (3) Summer 1989 229–41

Nelson, R. (1993) *National Innovation Systems: A comparative analysis* Oxford New York: OUP

Nelson, R. and Rosenberg, N. (1993) 'Technical Innovation and National Systems' Ch 1 in *National Innovation Systems: a Comparative Analysis* (ed.) R Nelson Oxford, New York: OUP pp. 3–22

Oakley, B. and Owen, K. (1990) *Alvey: Britain's Strategic Computing Initiative* Cambridge Mass: MIT Press

OECD (1986) *Innovation Policy* Paris: OECD

OECD (1989a) *Research Manpower: Managing Supply and Demand* Paris: OECD

OECD (1989b) *The Changing Role of Government Research Laboratories* Paris: OECD

OECD (1991) *Science and Technology Policy* Paris: OECD

OECD (1992) *Globalisation of Industrial Activities: Four Case Studies: Auto Parts, Chemicals, Construction and Semiconductors* Paris: OECD

OECD (1994) 'Accessing and Expanding the Science and Technology Knowledge Base' Working Group on Innovation and Technology Policy, OECD Directorate for Science, Technology and Industry, Committee for Scientific and Technological Policy, Paris: OECD

Olleros, F.-J. and MacDonald, R. J. (1988) 'Strategic Alliances: Managing complementarity to Capitalize on Emerging Technologies' *Technovation* 7 (1988) pp155–76

Pavitt, K. (1984) 'Sectoral Patterns of Technical Change: Towards a Taxonomy and a Theory' *Research Policy* 13 (1984) 343–73

Peck, J. (1996) *Work Place* New York: Guilford

Perrat, J. (1987) *Technologies, Externalités et Nouveaux* Rapport du Capital dans L'Espace Régional, Doctorat de 3e Cycle, Economie de la Production, Université Lumière Lyon II.

Perrin, J. C. (1974) *Le Développement Régional* Paris: P.U.F

POST (1993) *Science and Technology Agencies* London: Parliamentary Office of Science and Technology

Pratt, A. (1996) 'The emerging shape and form of innovation networks' Ch 7 in J. Simmie (ed.) *Innovation, Networks and Learning Regions?* London: Jessica Kingsley pp. 124–36

Preston, P. (1989) 'The Information Economy and the International Standard Industrial Classification (ISIC): Proposals for Updating the ISIC' *PICT Policy Research Papers* No. 6 February 1989

Preteceille, E. (1976) 'La Planification Urbaine. Les Contradictions de l'Urbanisation Capitaliste', *Economie et Politique*, No. 236, pp. 94–114

Quinn, T. (1994) 'Measure for measure' *Science and Public affairs* Summer 1994, pp. 42–8

Roobeek, A. J. M. (1990) *Beyond the Technology Race: an Analysis of Technology Policy in Seven Industrial Countries* Amsterdam: Elsevier

Rosenberg, N. (1990) 'Why Do Firms Do Basic Research (with their Own Money)?' *Research Policy* 19 pp. 165–74

Rothwell, R. (1992) 'Successful Industrial Innovation: Critical Factors for the 1990s' *R&D Management*, 22, 3 pp. 221–39

Ruberti, A. (1993) 'European Commission Guidelines for the Fourth Research and Technological Development Framework Programme 1994–98 XIII *Magazine* 1/93 No.10, 3 Brussels: DGXIII

Sanderson, M. (1989) 'Advances in Flow Measurement Techniques for Process Applications' Internal Document, Department of Fluid Engineering and Instrumentation, School of Mechanical Engineering, Cranfield: Cranfield Institute of Technology

Save British Science (SBS) (1994) *Public Investment in Research and Development* Oxford: SBS, p. 1

SBS (1995) 'Back to Square One' and 'Reduced to the Ranks' Autumn 1995 *Newsletter*, Oxford: SBS, p. 1

SBS (1998) 'At Last the CSR' *Newsletter* Winter 1998 Oxford: SBS, p. 1

Saxenian, A. (1991) 'The Origins and Dynamics of Production Networks in Silicon Valley' *Research Policy* 20 pp. 423–37

Sayer, A. and Walker, R. (1992) *The New Social Economy: Reworking the Division of Labour* Oxford: Blackwell

Scott A. (1988) *New Industrial Spaces* London: Pion

Scott A. (1998) *Regions and the World Economy* Oxford: Oxford University Press

Segal Quince and Partners (1985) *The Cambridge Phenomenon* Cambridge: Segal Quince and Partners

Segal Quince Wicksteed (1988) *Universities Enterprise and Local Economic Development: an Exploration of Links* London: HMSO

Schoenberger, E. (1989) 'New Models in Regional Change' Ch 5 in R. Peet and N. Thrift (eds) *New Models in Geography*, Vol. 1 London: Unwin Hyman pp. 115–41

Schwarz, M., Irvine, J., Martin, B., Pavitt, K. and Rothwell, R. (1982) 'The Assessment of Government Support for Industrial Research: Lessons from Norway' *R&D Management* 12, 4 pp. 155–65

Shotton, K. and Hillier, W. (1993) 'General Progress Report,' in Joint Framework for Information Technology *Annual Report* 1993 Swindon: SERC

Skoie, H. (1995) *Research and Technology Policies in the EC – Developments and Future Perspectives* Offprint from H. Skoie (ed.) *Science and Technology in the EU – General Development and Relation to the Nordic Countries Institute for Studies in Research and Higher Education*, Oslo, Report 8/95

Sternberg, R. (1996) 'Regional Institutional and Policy Frameworks and the Recent Evolution of RTD-intensive Enterprises in the Munich Region' Paper presented at the Networks, Collective Learning and RTD in Regionally Clustered High Tech Small and Medium Enterprises Project meeting, Nice 27–28 September 1996

Stoneman, P. (1987) *The Economic Analysis of Technology Policy* Oxford: Clarendon Press

Storper, M. (1993) 'Regional "Worlds" of Production: Learning and Innovation in the Technology Districts of France, Italy and the USA' *Regional Studies* 27 (5) 433–56

Storper, M. and Walker, R. (1989) *The Capitalist Imperative* Oxford: Blackwell

Storper, M. (1995) 'The Resurgence of Regional Economies, Then Years Later: The Region as a Nexus of Untraded Interdependencies' *European Urban and Regional Studies* 1995, 2 (3) pp. 191–221

Storper M. (1997) *The Regional World* New York: Guilford

Storper M. and Salais R. (1997) *Worlds of Production* Harvard: Harvard University Press

Swyngedouw, E. (1991) 'Homing in and Spacing out' Unpublished School of Geography, University of Oxford, Working Paper May 1991

Swyngedouw E. (1992) 'Territorial Organization and the Space/Technology Nexus' *Transactions of the Institute of British Geographers* New Series Vol. 17, No. 4, pp. 417–33

Swyngedouw, E. (1994) 'Exopolis and the Politics of Spectacular Re-development: Evaluating Ten Years of Urban-Regional Reconversion on Belgium's Deserted Coalfields' Paper presented at Urban Policy Evaluation Seminar, ESRC Seminar at University of Wales, Cardiff 7–8 September 1994

Swyngedouw E. (1997a) 'Excluding the Other: the Production of Scale and the Scaled Politics' Ch 13 in R. Lee and J. Wills (Eds) *Geographies of Economies* London: Edward Arnold pp. 167–76

Swyngedouw E. (1997b) 'Neither Global nor Local: "Glocalization" and the Politics of Scale' in K. Cox (ed.) *Spaces of Globalization: Reasserting the Power of the Local* pp. 137–66 New York/London: Guilford/Longman

Swyngedouw, E. (1998) 'Homing in and Spacing Out: Re-Configuring Scale', in H Gebhart (ed.) *Europa im Globalisieringsprozess von Wirtschaft und Gesellschaf,* Franz Steiner Verlag, Stuttgart, pp. 81–100.

Sutton, J. (forthcoming) *Technology and Market Structure: Theory and History* Cambridge, Mass: MIT Press

Tassey, G. (1991) 'The Functions of Technology Infrastructure in a Competitive Economy' *Research Policy* 20 pp. 345–61

Thrift, N. (1990) 'The Geography of International Economic Disorder' Ch 2 in R. J. Johnston and P. J. Taylor (eds) *A World in Crisis?* Second edition Oxford: Basil Blackwell pp. 16–78

Turok, I. (1993a) 'Inward Investment and Local Linkages: How Deeply Embedded is "Silicon Glen"?' *Regional Studies* 27, 5 pp. 401–18

Turok, I. (1993b) 'The Growth of an Indigenous Electronics Industry: Scottish Printed Circuit Boards' *Environment and Planning A* 25 pp. 1789–813

Van Dierdonck, R. Debackere, K. and Engelen, B. (1990) 'University-Industry Relationships: How Does the Belgian Academic Community Feel about It?' *Research Policy*, pp. 551–66

Van Geen, R. (1991) Chairman of the Belgian National Council for Science Policy, Personal communication

Van Kampen, T. G. M. (1989) '25 Years of Instrumentation' Report by WIB Chairman, the Hague, Netherlands: WIB

Van Tulder, R. (1991) 'Small Industrialised Countries and the Global Innovation Race: the Role of the State in the Netherlands, Belgium and Switzerland Ch 12 in U. Hilpert (ed.) *State policies and techno-industrial innovation* London: Routledge pp. 281–304

Walker, R. (1988a) 'The Dynamics of Value, Price and Profit' *Capital and Class* No. 35, pp. 147–81.

Walker, R. (1988b) 'The Geographical Organization of Production Systems' *Environment and Planning D: Society and Space* 7, pp. 377–408.

Walker, W. (1993) 'National Innovation Systems: Britain' Ch 5 in R. Nelson (ed.) *National Innovation Systems: a Comparative Analysis* New York, Oxford: OUP pp. 158–91

Webster, A. (1994) 'UK Government's White Paper (1993): a Critical Commentary on Measures of Exploitation of Scientific Research' *Technology Analysis and Strategic Management* 6 (2) pp. 189–201

Weinstein, O. (1992) 'High Technology and Flexibility' Ch 2 in P. Cooke, F. Moulaert, E. Swyngedouw, O. Weinsteinand P. Wells (eds) (1992) *Towards Global Localisation* London: UCL Press pp. 152–77

Whiston, T. (1992) 'Research and Higher Education: the UK Scene' Ch 2 in T. Whiston and R. L. Geiger (eds) *Research and Higher Education: the United Kingdom and the United States* Buckingham: SRHE and Open University Press pp. 18–23

Whittington, R. (1990) 'Changing Structures of R&D: from Centralisation to Fragmentation' Ch 8 in R Loveridge and M Pitt (eds) *The Strategic Management of Technological Innovation* London: John Wiley & Sons Ltd pp. 183–202

Yarrow, G. (1996) 'Discussion: the Problems of Multi-Regulation' in H. Lawton Smith and N. Woodward (eds) *Energy and Environment Regulation* Basingstoke: Macmillan pp. 314–16

Appendix

Project steering group members

Professor Roger Baker
Director
Gatsby Charitable Foundation

Mr Paul Bradstock
Director
The Oxford Trust

Dr David Budworth
Co-ordinator for the Joint Committee of the ESRC/SERC

Mrs June Clark
Director
Oxford University Industrial Liaison Office

Mr Ian Clasper
Senior Executive
GAMBICA

Mr Keith Dickson
Centre for Business and Management
Brunel University

Mr Nigel Gardner
Director
PICT

Dr Peter Jepson
Director of Research
Schlumberger Industries Flow Measurement

Professor Philip Hutchinson
School of Mechanical Engineering
Cranfield Institute of Technology

Mr Paul Hurst
NEDC Electronic Components Sector Group

Professor Mike Sanderson
Department of Fluid Engineering and Instrumentation
Cranfield Institute of Technology

Dr Marshall Stoneham FRS
Director of Research
AEA Industrial Technology
Harwell Laboratory

Index

Note: page numbers in italics denote tables or figures where these are separated from their textual reference